THE LAST
COLD PLACE

A Field Season Studying

Penguins in Antarctica

NAIRA DE GRACIA

SCRIBNER

NEW YORK   LONDON   TORONTO   SYDNEY   NEW DELHI

To all biological field technicians working far from home

CONTENTS

CONTENTS

CRÈCHE

FLEDGE

THE LAST
COLD PLACE

PROLOGUE

braced against the wind in the middle of a chinstrap penguin colony blanketing a rocky ridge. All around me penguins waddled through the colony or sat incubating their eggs on nests built from pebbles. The birds squabbled and crooned to one another. Some were in ecstatic display, flippers flung out from their bodies, heads pointing straight up, chests rising and falling in time with their screeching calls. The sound came from all directions and the noise was deafening. Penguin colonies are an assault on the senses: a cacophony of calls, a pungent fishy odor, and the penguins' short black-and-white bodies always in movement.

It was time for the annual nest census, which I was conducting with my coworker Matt. We tiptoed through the fray, laying bright ropes down between the nests to section them into countable pieces. Sitting on eggs, the penguins reached over to investigate the rope with their bills, tugging it and shaking it. By the time we finished, the ropes were brown and muddy with the thick penguin muck ubiquitous across all the colonies. We tried to discern the ropes' shape from uncontaminated patches of color and count the nests on a rusty tally-whacker, a metal yo-yo-size counter that kept track of the numbers every time my finger pressed a rusty button. A passing penguin mounted my boot and directed a flurry of flipper slaps at my calf, squawking its displeasure with characteristic chinstrap belligerence. If I'd looked down to nudge him off, I would have lost my concentration and had to count the section of nests again, so I let him go ahead and slap me. The sting kick-started my blood flow and pumped warmth to my increasingly numb feet.

It was a rare sunny day, early in my first season as a technician for a National Oceanographic and Atmospheric Administration (NOAA) ecosystem-monitoring field camp, and the snow-covered ground blazed with light. From my position at the top of the ridge, the small peninsula of Cape Shirreff was laid out before me, all rolling hills and ragged coastline. Since October, when I'd arrived, the snow on the hilltops had melted, exposing a dark earth where moss grew in small rusty patches and gray-green lichen awakened from a long winter dormancy. On the rocky beaches antarctic fur seals huffed and howled into the wind, in the full throes of their breeding season. I could see their shapes down below: the hulking bulls lording over their harems, the sleek females scattered on the rocks with squirmy pups tucked up against them. To the south was the Anguita Glacier, a great wall of ice, covering the land that connected the peninsula to the rest of the island. All other edges of the Cape bordered the sprawling expanse of the Southern Ocean.

The small thumb of land that was Cape Shirreff jutted out from Livingston Island, Antarctica. It was a mile and a half long, half as wide, and had been the entirety of my world for two months. Matt and I, along with three other field-workers, were living and working at the northern tip of the Antarctic Peninsula, just above the antarctic circle. South America, the nearest nonpolar continent, was about six hundred miles away. The nearest US base in Antarctica was some two hundred miles away. We had been dropped off by a ship two months ago, with all our gear and food, and would be picked up in three more. Our crew of five was the only human presence on this isolated promontory.

Our home was a one-room plywood hut that served as a living

living conditions
— compare to Shreve.

room, kitchen, office, and bedroom. We had no internet, no running water, limited electricity. We worked every single day, in all weather: snow, wind, rain, blizzards, gales, hail, sun. Christmas, New Year's, Halloween, weekends, full moon, new moon. We measured and counted, captured and released, tracked and took notes. My job, in essence, was to observe.

Our planet's southern ecosystem is changing, and there are few witnesses. The Western Antarctic Peninsula is undergoing the most dramatic regional changes in Antarctica, with one of the fastest warming rates in the world. Since 1950, average winter temperatures have increased by 7°C (five times the global average). The onset of sea ice in late autumn comes later year by year, decreasing the number of annual days of ice and the amount of ice overall. In Antarctica, every life-form is attuned to the yearly cycle of ice. For antarctic krill, the species on which this whole ecosystem depends, ice is a nursery. The decrease in sea ice is causing cascading changes across the antarctic ecosystem— some of which we are only just beginning to discover.

Long-term monitoring projects such as the one at Cape Shirreff allow us to see these changes play out among Antarctica's top predators. Seals and penguins are indicator species: changes in their reproductive success and population can tell us what is happening with their food source. The information we gather at the Cape is ultimately brought before CCAMLR, the Commission for the Conservation of Antarctic Marine Living Resources, the international body that sets limits to the global krill-fishing industry. NOAA administered the monitoring program on a national level as it fell under its oceanic purview.

Every figure about antarctic marine species is the result of an enormous output of time and labor. For every single dot on a graph scientists present to clean-cut diplomats and policy makers, there is a grimy fieldworker like me, stationed on an isolated island, surrounded by penguins, covered in penguin muck and smelling like fermented shrimp, writing down metrics and surveys in an equally grimy field notebook.

For every long-term population trend reported in a journal article there are decades of field biologists standing in wind and snow, monitoring penguins or seals, hitting tally-whackers with numb fingers, far from family and friends and anything resembling human civilization. Our lives are tied to the weather, the season, and the wildlife itself.

It's not glamorous work. You're dropped on this frigid island with four other people and no privacy. Your body is buffeted by the elements, your mind strains under the work's demands, your heart is rubbed raw with beauty. You live among wild things in a wild place, stand on a stark island facing your own stark nature. There are no shops, no roads, no TVs, no trails, no distractions from the machinations of your own mind. Just a handful of lonely shelters, your crew, the wind, the rocks, and the penguins.

Yet, it is a joy to work at Cape Shirreff, for reasons that, to field biologists, seem self-evident. But personal fulfillment does not fund research. It is the curse of a biologist to justify the existence of their study species to the world. Beneath the grant proposals, paper introductions, and presentations must lie proof that the species under scientific scrutiny have immediate value to our society.

This is particularly complicated for ecosystems that remain far from human experience. Only recently have humans populated the barren shores of this continent, and even then, ever so sparsely. What of ecosystems so remote that few will ever encounter them? In the two five-month field seasons I was there, I kept trying to articulate to myself the value of other species to human lives, beyond their scientific role as indicators for the wider ecosystem's health. Why should we work so hard to maintain biodiversity in a place so remote? Why, exactly, should we care about penguins?

AUTHOR'S GOAL

LAY

1.

Mid-October

On the morning of October 27, 2016, our inflatable dinghy cut through frigid waters toward the northern shores of Livingston Island. It was the first land I'd seen since we passed the last southern islands of South America five days earlier. The Zodiac carried the Cape Shirreff crew—Matt, Sam, Whitney, Mike, and me—plus a Zodiac operator from the ship and a couple of people headed to the antarctic base Palmer Station, on Anvers Island, who came to help us unload. We were all swaddled tight in bright orange float coats. I sat on the side of the Zodiac gripping wet rope and watching the bright metal hull of the ship shrink as we rode the billowing dark swells toward the island's coast.

We'd just crossed the world's stormiest stretch of ocean—the Drake Passage, the shortest crossing of the Southern Ocean from South America to the Antarctic Peninsula. At this latitude there are only a few scattered islands—no landmass significant enough to obstruct or slow down the antarctic circumpolar current, which whips around the globe unimpeded, speeding across the Southern Ocean with a ferocity unmatched anywhere else on earth. On a broad orange ship, we'd climbed the rolling hills of ocean swells, pitching forward and back, forward and back.

After five days of travel, we were finally on the Zodiac, pulling up onto a sheltered beach. The fog obscured everything except the rocky shore and our camp, which sat on a slight rise two hundred yards away. The huts looked like matchboxes scattered on a blanket of white, dropped from a giant's pocket. I could barely discern the vague outlines of bigger hills. Cape Shirreff is one of two ice-free peninsulas on Livingston Island, 88 percent of which is covered by an ice cap. The island was just offset from the very northern tip of the Antarctic Peninsula, with nothing but open ocean between us and the southern edge of Chile, about five hundred and fifty miles away. King George Island, our more populated neighbor, was fifty miles east, the former home of the monitoring program and where many other bases still operated: South Korea, Chile, Russia, Argentina, Brazil, Poland, and Germany all have a presence there. Livingston, in comparison, hosted the huts of our American camp on a northern peninsula, with two Chilean research huts nearby and small seasonal Spanish and Bulgarian bases on its southern coast. While the island was about forty miles long and fifteen miles wide, my world would exist solely on a tiny peninsula attached to the northern finger.

I breathed in the biting air as I wrestled off my float jacket, relishing the stability of the ground beneath me. I still felt the world swaying: forward and back, forward and back. The air temperatures hovered around $-1°C$, and the beach was littered with huge chunks of ice. At the high-tide line, the shelf of snow that covered the island was almost as high as my head.

The crew was composed of two seabird technicians and two seal technicians, one new and one returning on each team, plus one NOAA researcher from San Diego, where the Antarctic Marine Living Resources research program was headquartered, as our crew lead. When we accepted the job, we signed on for two seasons: one to learn and one to teach. Matt, my fellow seabird technician, was returning for his second season, joined by Whitney, the lead seal tech. Sam and I were new, and Mike, the NOAA researcher, rounded out the crew. As Matt and Whitney well knew, our first task was to dig steps into the wall of snow that rose from the rocks. Sam and I followed their lead in all things. Shovels out, we carved.

The world's coldest and most remote continent sits like a blank white footer at the bottom of our maps, stretched to fit the flat cartography of a northern-centered worldview. But when mapped from a south pole perspective, Antarctica is close to round. East Antarctica—"East" in reference to Europe, of course—is the continent's biggest component part, fanning out from the south pole like an ear. East Antarctica holds the continent's thickest ice shelves and vast, snowless dry plains, both largely lifeless save for microbes and the occasional Weddell seal that wanders too far inland. The Transantarctic Mountains mark the boundary between East and West Antarctica. West Antarctica faces the Pacific Ocean, with two massive waterborne ice shelves on either side. Sticking out like a curved thumb between the ice shelves is the Antarctic Peninsula, the only part of the continent that reaches north of the antarctic circle.

Life on the continent thrives on the shore, where marine species use land to nap or breed, but there are no year-round terrestrial animals. The waters that fringe the continent teem with seals, whales, krill, fish, penguins, and strange and unique seafloor creatures, such as bright red sea urchins with vicious white spines and giant feathered sea stars.

Most of antarctic biodiversity depends on the rich waters of the Southern Ocean. The 1959 Antarctic Treaty, however, only assured protection for the land itself. The treaty was signed by the twelve countries—Argentina, Australia, Belgium, Chile, France, Japan, New Zealand, Norway, South Africa, the Soviet Union, the United Kingdom, and the United States—that had been active in Antarctica during the international polar years between 1882–83 and 1932–33. The international polar years were organized to coordinate data collection on Antarctica and the upper atmosphere. In the years following the signing of the treaty, as mariners discovered the vast untapped "resource" of antarctic krill that bred and thrived in this remote marine environment, commercial interest in the Southern Ocean grew. Beginning in the 1970s, ships from the USSR began fishing for krill, soon joined by ships from Japan, South Korea, China, and Norway.

Antarctic krill has the most biomass of any animal species on earth. If massed together, antarctic krill would weigh twice as much as the entire human population. Vast swarms spanning up to twelve miles—the largest animal aggregation in the world, another record—are scattered across the Southern Ocean in a patchy distribution that helps them avoid predators. Krill represent an abundance that challenges any reference we might have for the word, a staggering organic expanse coexisting in unified masses, suspended like a cloud of dust motes in cold and unforgiving waters. Krill is the keystone species of the Southern Ocean's food web, holding the whole ecosystem together. Seals, penguins, whales, and many fish all depend directly or indirectly on krill for their survival.

As more countries began fishing for krill, the scientific committee that formed part of the Antarctic Treaty raised concerns that the krill fishery would impact the whole antarctic ecosystem. At the 1972 meeting of Antarctic Treaty Consultative Parties, representing the treaty's signatories, a resolution was approved to invest in scientific study of the Southern Ocean and establish a system under which it could be protected. This led to years of working groups that developed a conservation system for "Antarctic marine living resources" as well as a huge research initiative to understand the basics of the Southern Ocean ecosystem. Emerging from these efforts, the Commission for the Conservation of Antarctic Marine Living Resources (CCAMLR, pronounced *kam-lar*) began operating in 1982. The twenty-five member countries of CCAMLR, including the United States, contribute to its budget, send representatives to meet yearly to make decisions, and participate in antarctic research.

The Antarctic Treaty system is the overarching international governance structure for research and regulation on the continent, something like a continent-specific UN. CCAMLR is a convention that forms part of the "treaty system" and is designed to address a specific antarctic issue—the conservation of marine life around Antarctica. CCAMLR takes an ecosystem approach to regulating the krill fishery, which involves establishing long-term ecosystem-monitoring programs throughout the continent, run by the parties that signed the convention. The United States contributes to ecosystem-monitoring research as a member of CCAMLR, and NOAA runs these research programs within the United States, as directed by a 1984 congressional mandate called the Antarctic Marine Living Resources Convention Act.

The two goals of CCAMLR's ecosystem-monitoring program are to detect and record changes in the marine ecosystem around Antarctica and to distinguish between changes due to the fishery and due to environmental variability. A blunt approach to regulating the fishery would entail studying the target species only: the population and growth rates of krill, and how much can be harvested without decimating numbers. But an ecosystem-monitoring approach sets fishing limits that also account for krill's role in the Southern Ocean marine ecosystem and is designed to avoid significant impacts on other species due to the harvesting of krill.

Fishing is not the only force acting upon the Antarctic food web. Climate change is poised to have a big impact on the ecosystem of the Southern Ocean, much of which depends on the yearly cycle of sea ice. While the long-term monitoring program was initiated to measure the impact of the fishery, climate change has become a key focus of the research conducted at CCAMLR's ecosystem-monitoring sites.

With so many forces at play in these polar waters, how to distinguish between the effects of warming and the effects of commercial fishery? How does climate change impact krill populations in the Southern Ocean as a whole? Why, and how much, do just a few degrees matter in this ecosystem? How is CCAMLR to set precautionary fishing limits that account for both the impact of climate change and the fishery? Monitoring stations across the continent and surrounding waters collect the data that can lead to answers.

The five months at Cape Shirreff dovetailed with the summer breeding season of our target species: chinstrap penguins, gentoo penguins, and antarctic fur seals. Most of the methods we used were standard ecosystem-monitoring protocols developed by CCAMLR to ensure that data across projects could be compared. To monitor the penguins, we'd be documenting nest counts, adult survival, adult weight, egg weight, egg lay dates, chick hatch dates, chick growth rates, chick survival, and the composition of penguin diets. We'd attach data loggers to penguins to measure the duration of their foraging trips, how deep they had to dive to find food, and where they found it. While we were there, we also monitored the reproductive success of skuas, predatory seabirds that feed on penguin eggs and chicks during the breeding season. Sam and Whitney would track the breeding of antarctic fur seals and would keep tabs on all the seals that hauled out on the peninsula.

The same ship that plucked us from the southern tip of South America would pick us up again in March, after all the penguin chicks had fledged and the fur seal puppies had weaned off their mother's milk, prepared for a long winter in a turbulent sea.

Two Zodiacs shuttled gear from the ship to the shore: all the food, equipment, and personal bags we'd need for the next five months. The gear piled up on the beach. Some gear was stashed in the large plastic crates that had been buried in the snow. The crates protected gear while it was staged on the beach. After hauling plastic totes and drybags onshore and lugging them up toward land, Sam and I headed to camp for the first time, laden with boxes of food. On our way, we spotted a penguin walking what we later learned was a well-worn trail leading from the beaches south of camp to the penguin colonies. We pointed at it and exclaimed, "Look! A penguin!" I was still so green to Antarctica I didn't even know what kind of penguin it was. Matt, freckled and with his characteristic bushy ginger beard, sidled up and informed us that it was a gentoo. Sam and I just stared at it, beaming and wide-eyed.

While Sam and I were near strangers, Matt and I had met three years before on a windswept island in the Bering Sea, somewhere between Alaska and Siberia. He had tutored me in banding wriggling murres we caught on steep cliffs, in the steady patience required to catch and work with seabirds, in the determination necessary to get things done in inclement weather and acute discomfort. Along with everything he had to teach me came the curious mind, ironic humor, and sense of adventure that made us close friends. At thirty-five, he'd been hopping around on the field circuit for a decade longer than I had. When I first met him in my nascent field career, I had been awed by his decade of islands and birds and fieldwork, thinking that this was exactly what I wanted to be doing: moving from one island to the next, always living in bunks or tents, largely working outside in far-flung places that numbered higher in birds than people. *Intro to her study in seabirds*

I'd heard about the Cape Shirreff program as many field techs do: from other field techs, part of the field-technician world and culture I'd first joined during college summers, working on seabird-monitoring islands in Alaska. My first window into fieldwork was a stint in northern Maine. I got to stay on a seabird-nesting island for ten days, and I was sold. For my second full field season, I applied for a student program with the Fish and Wildlife Service and headed to southeast Alaska, a temperate rain forest bursting with storm petrels. There, I worked with someone who had cut her teeth down on King George Island, which neighbors Livingston Island. I heard the stories of cold hands and violent storms, of a rock in the middle of the Southern Ocean where penguins were a part of daily life. Back at university, I worked for a professor who'd also done many years down at King George, and I heard more stories of mold and the pungent aroma of penguin. I took to fieldwork with fanatic zeal and set my sights on the antarctic.

For a field technician, Antarctica was the ultimate season: ultimate remoteness, ultimate species, ultimate immersion. I'd wanted to go since the very first story graced my ears. To apply, one had to chase up one of the lead researcher's contact information through friends or colleagues. The job was not advertised. To get a spot, it helped enormously to have someone known to the program who could vouch for you—who'd worked with you in the field. I had the credentials, but when I applied, I also had Matt's backing, which meant a lot from the person whom I'd be working with the most.

It had been two years since I'd seen Matt, back when we'd road tripped up the West Coast to his new home in Alaska. I flew back to California once we reached Juneau—I'd just graduated college and was about to start the next year working as a farm manager at my university. Aching for fieldwork, I ran off to work on another island once my year was up: Midway Atoll, a speck in the Pacific Ocean halfway between California and Japan. From Midway, I'd written Matt long emails about the albatross that covered every square inch of the island, about learning all the native plants and restoring sandy habitat. He wrote me from Cape Shirreff, describing the hut, the wind, the landscape, and penguin mannerisms.

Heading into a small field camp for five months, where getting along with one's crewmates could make or break the season, I knew I could count on him and that was a balm to my nerves. Matt was like a boulder held steady in the middle of a rushing river. The contrast between my incessant forward drive and Matt's thoughtful progress is one of the things that make us a good team.

Sam, Matt, and I headed toward the snow-covered cluster of huts we called camp. The largest of the huts was our main living space and the first priority for opening. Mike and Whitney had been busy uncovering all the essential doors and propane hookups along the outside. After the door had been dug out and its cover removed, Whitney pried it open and stepped inside.

Main hut, stark and dim, would be our bedroom, kitchen, dining room, office, and living room. The walls were painted white—what you could see of them, at least, behind the shelves and maps and pictures and gear. No space was left unused. Everything was bagged in plastic for overwintering in March the year before, an attempt to stave off the mold. And I mean everything: folding chairs, all the kitchen appliances and food, all electronics, dishes, bowls, pans. Unlike Palmer Station, a year-round research station, Cape Shirreff is a seasonal monitoring camp, meaning it only operates in the summer months, during the wildlife breeding season.

It was hard to see a shred of coziness and livability in this collection of moldy wood panels and bagged utensils. I was reminded of all the times, growing up, that I walked into a space for the first time knowing it would be my home for the next however many years. Every time I moved to a new place, I'd project myself into the walls, measuring the dimensions of my future, trying to imagine my life in the empty spaces. The main living hut was standard fare for a field camp: bunks, a table, a kitchen, a desk area for data, and little else.

Mike introduced every hut and area of camp with a story from his decades of field seasons. He'd had a hand in naming almost everything, which meant a lot of time-honored jokes were passed down to us with the rest of the work. Before working at the Cape, Mike had monitored a small population of fur seals on Seal Island, a tiny rock in subantarctic waters, from 1986 to 1992. In 1995, the director of NOAA's Antarctic Ecosystem Research Division asked him to establish a long-term monitoring program at Cape Shirreff based on CCAMLR research protocols. Chilean researchers had been going down to the Cape as early as 1991 to conduct fur seal counts, and the Cape, with both breeding seal populations and penguin populations, was an ideal site to estab-

lish an ecosystem-monitoring camp. While the legislated purpose of
CCAMLR's monitoring program was to inform fishing regulations by
tracking indicator species, scientists were beginning to see warnings of
growing levels of greenhouse gases in the atmosphere, and it became
increasingly clear to the program leads, Mike and his seabird counterpart, Wayne Trivelpiece, that the effect of climate change on this remote
ecosystem would be a critical part of their research.

The bare bones of the camp were built in the austral summer of
1996–97—what in the northern hemisphere would be the winter—and
it was a bafflingly complicated undertaking, as everything that makes
up the physical structure of camp was brought in via inflatable dinghy
and off-loaded on a rocky beach. With the climate already shifting conditions in the Antarctic Peninsula, Mike, Wayne, and their hired crews
rushed to establish a baseline against which future changes could be
measured.

Besides main hut, camp also featured an attached workshop and
small lab for preparing and sorting fur seal samples such as scat and
blood: a short counter a yard and a half across, stacks of shelves with
vials, tubes, and sampling equipment, and a microscope perched in
the corner. The "stay-wet room," initially intended as a room for drying
things, adjoined main hut and was the shower and laundry space. A separate supply hut a few meters from main hut was our pantry and had an
extra small room in the back, which was typically used for the research
leads such as Mike to sleep in blissful privacy (the "old fart's room"). A
wooden outhouse (two seats and two buckets: one for solids, one for
liquids) stood off on its own, near a small garage for the utility task vehicle (UTV). Whitney and Mike had fired up the UTV, complete with
snow treads, and Matt was driving it back and forth from the beach to
camp, pulling sleds piled high. Matt assigned me to the "freshies room,"
a space joined to the side of main hut, where our vegetable crates, chest
freezer, and other food that required refrigeration would be stored. Refrigeration, in this case, simply meant that the space was unheated. The

crew directed all vegetable and fresh-food boxes to me, and I frantically packed peppers and cabbages and potatoes in crates, stacked cheese on a shelf, and threw seafood, meat, and other frozen foods into the chest freezer, which was not plugged into anything but rather would be kept cold by ambient temperatures and all the frozen things we would stuff into it.

Settling into camp

Piles of gear lay everywhere, between all the buildings, behind camp, in front of camp, on snow that covered a theoretical deck. Food, sample-processing equipment, materials to maintain the physical structure of the huts, capture equipment, clothes: It seemed as if we'd never unpack it all. After many Zodiac runs, a mate hailed us on the radio system Mike had revived from its winter dormancy. The ship wouldn't leave until our communication system was up and running. We'd also have a satellite phone and a satellite email service on a single computer, to be set up in the first days of camp.

Once we confirmed that all was well, I watched the *Laurence M. Gould* pull away and with it the last link that tethered me to the rest of the world. The stillness of the island settled over me. I took a deep breath. It was just the five of us now. I was nervous, exhausted, and grinning madly.

Finally on the shores of the white continent, I felt like I stood on the edge of a deep, vast chasm that gaped between the present and the March pickup at the end of the season. I was eager for all of it—the discomfort, the wind, the chores, the fieldwork itself—ready for everything I was about to feel and experience that I couldn't yet imagine. In past seasons on other islands, my expectations had been upended by the social and cultural world that developed within a small crew, working mostly alone for long hours in wet, cold, or windy conditions. Antarctica was my most remote field season to date. Though the parameters of the work were clear, our environment was unpredictable—everything we did depended on both volatile weather and volatile wildlife. Every day could bring something unexpected.

We spent hours hauling gear up to camp, slogging up the hill through the snow. By evening, essential foods and boxes had been put away, everything on the beach was secured and organized, and all essential entry points had been dug out. We settled into the cabin. It exuded a musty, moldy smell that told of a long, wet, cold winter. All our cleaning supplies were frozen solid and therefore useless. Outside the snow blew horizontally, but I was too flushed from activity to be cold.

Matt, Sam, Whitney, and I would sleep in the main hut bunks, and Mike in the back of the supply hut. Having claimed our beds—Matt and Whitney had first dibs and quickly beelined for the ones they'd had last year; Sam and I took the other bunk bed, with me in the top bunk—we made our first dinner in camp.

The time-honored tradition was to buy empanadas in Punta Arenas and throw them in the oven to heat on the first day, to avoid unpacking the whole kitchen to cook dinner. Mike couldn't seem to get the oven to work, so we ate them cold, sitting in folding chairs at the table, drybags scattered on the floor all around us, kitchen half-unpacked, outdoor gear still on. Whitney turned on the recently revived propane heater, waking up the ice inside a blue five-hundred-gallon water barrel from its long winter hibernation. As the cabin warmed for the first time in seven months, the ice hissed and crackled and growled, and we laughed and laughed because it sounded as if something were trapped in there, and because we were so tired and aflutter we wouldn't be able to stop giggling even if the ice fell silent.

Mike, Matt, and Whitney had all worked at the Cape together the previous year, and all of us except for Sam had been bumping around the seabird and seal field-job circuit for years. Long-term ecosystem-monitoring studies need people on the ground, stationed in remote field camps, documenting the populations and reproductive successes of species of interest. Many folks that study ecosystem-level biology look to get some fieldwork experience before moving up as researchers or natural-resource managers, building off degrees in biology, ecology, natural-resource management, or wildlife science. Field technicians are a mixed bag, and while there is no majority career path, a common trajectory is to move from fieldwork to jobs as a crew lead, then to grad school, and eventually into a research-coordinator and camp-supervisor role. A few rare ones tackled a PhD and ended up higher in refuge-career levels, managing several camps and steering whole research programs. Some field techs had no desire to move up the career ladder and simply fit into fieldwork like a hermit crab into its shell, hopping from one remote, seasonal data-collecting job to another.

When I first started doing fieldwork in college, my guiding professor, who'd been around the circuit many times herself, told me that at a certain point I'd want to ask my own questions and design my own studies, rather than collect data for someone else's research. My expectation had always been that I'd go to grad school, carry out my own research project, maybe get a PhD, work as a researcher at an ecosystem-monitoring program, and eventually run camps like those in which I worked. After five years, though, I still loved being someone else's hands on the ground—I felt as if I had all the fun and less of the responsibility.

In the evening, as the dust settled, I took a moment to absorb the shape of my new home. I stood by the propane heater, hands outstretched to catch the warmth, staring out the window to the snow, beach, and ocean beyond. So much technology was necessary to get us out to camp—the massive, hulking ship, the Zodiacs, the UTV, the vast networks of governance and regulation coordinating across international boundaries—but once we were here, life was quite primitive. A small band of people huddling around a fire, just as our ancestors once had.

Antarctica is the epitome of everything that is, in the Western worldview, wild and remote. Two years out of college and beginning my fifth field posting, I was ever chasing the high of a far-flung island, of rocky beaches and birds, embracing a lifestyle that indulged an innate restlessness I've never quite been able to shake. The island I'd worked on before had been fertile ground for my wandering mind—ever since Alaska, I'd tried to unravel the ways social and economic paradigms shaped the intimate relationship between me and the ecosystem around me. Antarctica was the ultimate proving ground; not a final destination, but an experience that might lead me toward the kind of existential revelation I always felt was waiting on the next island. In a pursuit to understand ecosystems, maybe I could understand everything. Maybe I could even understand myself.

I'd already given the crew my usual "Where are you from?" spiel: My parents are journalists; my mother's family is American but she grew up in Mexico, where her family has lived for two generations; my father is from Spain (the only uncomplicated part of my heritage); my parents met in Panama; I was born in California, then we moved to Spain, Mexico, Chile, Argentina, South Africa, and Egypt; and I moved back to California for college. When I was young, I crossed so much distance so often that it became a simple fact of my life: I was a foreigner everywhere, and that distance often came between me and the world around me. When I started working on remote islands, distance served to keep the rest of the world away, and in the absence of society, I found a deep, intimate relationship with the biome in which I lived, a closeness, a resonance. The distance between me and my host ecosystem would shrink, become a sliver, and sometimes disappear altogether.

Though I'd eagerly been following a trail of islands—from Maine, to Alaska, to Hawaii—I wasn't yet sure where they would lead. The milestones years down the track were like a mirage, barely there, unfocused. Hastily sketched from what others had done after walking this path—not exactly mine and not exactly front of mind. Warming my fingers by the fire and looking out at the light playing on snow, I was too wrapped up absorbing the antarctic island that would be my home to imagine what might lie beyond it.

2.

Late October

After a few days, we'd set up camp enough that Matt decided it was time for us to venture to the penguin colonies up north, where the two of us would work every day of the season. At the tail end of the antarctic spring, the landscape would be easing into summer and the plethora of marine life that lived off the continent's shores would return. The antarctic summer would be abundant in light, but not so much warmth: we'd go from average temperatures hovering around 0°C to one or maybe two degrees above zero in the middle of summer. Matt had regaled me with stories of the previous year's unusually violent weather, so I was largely preparing myself for wind and storm.

While the seals clustered on all the beaches, the penguins bred in colonies in a specific area in the northern edge of the Cape. We moved past the sloping ground just by camp, to the hilly land beyond—a slight uphill slant that had me panting as I heaved my snowshoes through the snow. The Cape was all gentle rolling hills and dark rocky beaches. We headed down to a beach where fur seals fought, up over a hill and across a ridge just above another beach, until we turned the corner on a slight rise to the colonies.

The land sloped toward shore, the snow dotted with penguin colonies, streaks of their bright pink excrement ringing every group. Penguin sounds and smells suffused the air, and I stared, delighted.

Penguins waddled along from sea to colony, peering at us as if we were just another, taller kind of penguin. The chinstraps (or chinnies) are a cantankerous species, stockier, with a namesake strip of black below their faces. Gentoo and chinstrap penguins bred in the same area, on a number of rocky outcrops a mile north from camp, close to the northern beaches and the edge of the island. The penguins bunched together as colonies on small rises, covering about a third of the area in twenty colonies total. Each colony could have somewhere between twenty and five hundred breeding pairs. Four of the colonies were exclusively gentoo, ten chinstrap, and six had both species nesting together.

The gentoo penguins were larger, with bright orange bills and elegant white headbands. Mated pairs bowed to each other courteously and some were already arranging pebbles into neat bowl-shaped nests. Antarctica has no nest-ready plants to speak of, just patches of moss or lichen and a sparse and particularly hardy species of grass. Pebbles, however, are everywhere, and penguins use them to build hard but well-drained nests, an essential quality in these wet and stormy lands.

Both penguin species were part of the genus *Pygoscelis*— "brush-tailed" penguins. The third and last member of the genus is the Adélie penguin, which did not nest on Cape Shirreff but did nest in islands nearby, including King George. Chinstrap penguins spend their winters in the open ocean and seemed disgruntled to be crammed into a chaotic colony. Like albatross and all other pelagic seabirds—such as murres, kittiwakes, puffins, shearwaters, auklets—chinstrap penguins feed and sleep at sea, bobbing with the waves as they sit on the surface of the water like a duck. Chinstraps journey long distances in winter, moving west along the continental shelf. They are ice-avoidant, meaning that they prefer open waters and generally steer clear of sea ice. Gentoo penguins are not migratory and tend to be more localized in the winter. They stay in the waters near their breeding colonies and sometimes hang out on land. Adélies prefer ice and spend their winters resting and hunting from the pack ice that grows from the shore outward.

The penguins that thronged around me had survived the dark antarctic winter, made their way back to shore, found a mate, and were already busy staking out nesting spots and crooning to their partners. Matt saw breeding pairs, could distinguish between sexes and discern where they were in their breeding season, but all I saw was a sea of penguins. Early in the season, they all looked the same to me.

The first task in penguin lands was to open the skua shack, named for the predatory seabirds that ever circle the colonies. This small hut served as the home base for all penguin work and for the penguin crew—just two people for five months of the year. The shack stood on a snowy hillside, weather-beaten walls still standing at the end of yet another long and lonely winter. It was built on a raised platform in an attempt to lift it clear of the snow and the muck. Besides being penguin headquarters, the shack was our emergency shelter if main hut burned down or was destroyed in some other catastrophe. Like on most buildings at the Cape, the paint was peeling due to the extreme wet and wind. The west-facing wall was almost completely stripped. Seal skulls were nailed every half yard around the outside of the boards of the deck that circled the shack, like a warning or a testament. While Matt hacked away with a pick at the ice frozen around the door cover, I unscrewed the wing nuts on all the window covers. When the ice yielded, we wrestled off the plywood and stepped inside.

It smelled like main camp—mold, mud, and wet wood—but unlike main camp, there was a distinctly fishy undertone. Two skua wings were nailed over the door. A couple of wooden bunks lay along one wall, followed by a modest kitchen counter. On the other side was a row of windows, a work counter, and a wooden table with two office chairs.

No one had been in the shack since the end of the previous season, seven months earlier, and the chairs sported a coat of mold, filaments of delicate green hyphae growing from the synthetic blue fabric into the humid air. The mold elsewhere was less showy, just a dimly green sheen, but we wiped it all off regardless.

Below the counter and linked to the two-burner camping stove was a valve for propane, which connected to a full propane tank outside that had been hauled here at the end of the last season. Matt turned the propane valve and showed me where the big red button was to turn on the load for the solar array. Five solar panels outside were mounted below the windows and would power our skua shack data laptop. I took stock of the snacks that had overwintered (granola bars, nuts, stale crackers, frozen for months and recently thawed) and plopped into my chair to test its comfort. Out the window, I saw for the first time the view that would characterize my days at the shack: the penguin colonies on land, the beaches just beyond them, and a narrow rocky outcrop that divided my view of the ocean in two. I could get used to this, I thought.

"Comfy?" Matt asked.

"Yeah." I sighed.

"Good. Because you will spend *a lot* of time sitting in that chair."

On the wall above the windows a line of framed pictures featured the two penguin technicians of past seasons, standing outside among penguins. The first was from 1996, and the last somewhere in the early 2000s, when the tradition petered out because, I suspect, no wall space was left. The people in the pictures wore the same field-issue gear Matt and I now wore, sported the same noses rubbed raw by wind, and were surrounded by generations after generations of monitored penguins. The rocks, like the penguins, hadn't aged a bit, nor, it seemed, had the shack itself. The amenities at the skua shack were so rudimentary that it probably looked much the same as it did twenty years ago, untouched and unbothered by the movements of the rest of the world, solid and functional with its basic four walls and two-burner camping stove. I

Character dev.

wondered what had become of all those field techs on the wall, frozen
in time. What had they done next? Where were they now?

The sense of history that always permeates long-term monitoring
projects never ceases to humble me. In 1997, the first full year of penguin monitoring on Cape Shirreff, I was five years old, learning to write
cursive in Santiago, Chile, while a few thousand miles south two people worked in this very hut, holding notebooks similar to the one now
stashed in my bag, looking for banded penguins. In 2000, I was eight,
living in Argentina, reading my first short chapter books and wearing outfits of only one color, while the two people a few pictures over
walked through the wind, lifting penguin tails to check for a second
egg. In 2010, I was in my last year of high school in Cairo, Egypt, and before I'd even learned what biology fieldwork was, before I'd even known
working on islands was possible, two more field techs knocked the
snow off their boots before stepping into the shack. While I drank one
last beer in the dusty, desert city that was my home, someone learned
how to catch a penguin.

Matt and I were taking up exactly the same work demanded by protocols set up long ago: track the penguins, as they went from a small pile
of pebbles to fledging chicks, throughout their breeding season by following a subsample of nests, and noting every significant date in tables
drawn in small waterproof notebooks. We'd record the number of eggs
and date they were laid, the number of chicks hatched and when they
hatched, and the chicks' growth and how many survived to adulthood
and fledged into the ocean. Keeping data-collecting methods consistent is essential to building a long-term data set and identifying trends.
Parameters such as timing of the first egg, growth rate of chicks, adult
foraging-trip duration, and proportion of chick survival were all pieces
of the puzzle, pointing to shifts in a larger picture.

We were far from the only camp undertaking this work. Cape Shirreff was one of thirty-two CCAMLR ecosystem-monitoring sites, and
one of two run by the United States. The other of the two was on King

George Island, and the penguin research lead, Jefferson, who usually worked in San Diego, went to King George Island for a few weeks every other year and ran an abbreviated version of Cape Shirreff's monitoring, largely pivoting around diet samples and device deployments. The other ecosystem-monitoring sites around the continent were run by other signatories of CCAMLR, including Argentina, Australia, Uruguay, Ukraine, New Zealand, and Korea. All parties convened once a year in Hobart, Australia, to share significant findings in their research and address subsequent changes in fishing regulations in the Southern Ocean.

On the ground, fieldwork was more like a lifestyle than a job. Fieldwork had taught me the patience needed to peer into an ecosystem and try to discern how it worked. After my sophomore year of college, I worked on an island in southeast Alaska called St. Lazaria, a volcanic rock covered in salmonberries and Sitka spruces. My days were spent sitting on a cliff edge staring through a scope at another cliff edge where all the murres nested. The adults guarded the eggs closely enough that I had to stare at them for hours to catch movement that revealed an egg or chick. As I was ever rushing from one thing to the other, be it countries or projects or books, this kind of observation required patience I didn't think I had. It forced me to slow down. It forced me to really look.

Sometimes, I'd stop observing the nests I was assigned to monitor and look around me with the same attention. I watched a gull ride the wind that flew up the cliff, wings spread, surfing with effortless care and grace the gusts that battered me so endlessly. I'd become absorbed by the delicate, transparent membranes of the moss clinging on the rocks or watch seaweed dance in the tide.

Other moments were more tangibly spectacular. Once, early in the season, the crew and I went out in the evening to watch the storm petrels coming in. The birds lived in burrows in the ground, hunting in the day and returning at night. We were usually in bed by then, lights off lest we disorient them. But that one night we sat on the moist ground for hours, the island quiet as the sun faded, and I dozed off. When I

woke, thousands of flapping wings filled the sky, storm petrels streaming onto land, stepping through the grass, scuttling into their burrows
to feed hungry chicks. They called from the air, from the ground, and
from beneath it. We were completely surrounded. The birds bumped
into us on their way home, climbing over our legs and hands as if we
were more fallen branches to be navigated. In the burrows, the storm
petrels purred loudly, a whirring that when combined seemed to rise
from the ground as if the soil itself had awoken.

Enjoying the sight of a penguin colony, or a seal harem, or a cloud of
storm petrels was one thing, but staring into the colonies at individuals
every day for months was entirely another, tasks that Mike and other
scientists did year after year, monitoring the wildlife to understand how
their ecosystem might be changing. Wonder is the fire behind so many
of the best scientists I know, and behind so many moments of intuition,
research ideas, and protocols. Science is not totally objective or apolit-
ical, but it can be as beautiful as the ecosystems it studies. Beneath the
fanfare of academics and opaque, scientific language lies the simple joy
of living on this earth and the urge to understand it. The seed of wonder
that had been planted in me as a kid, climbing trees, peering at flowers, looking for birds, had grown by orders of magnitude in ecosystem-
monitoring camps, when I was forced to look, really look, at the biomes
that made up my world.

While Matt, I, and the field technicians pictured on the skua shack
wall were temporary presences in the colonies, lead researchers such as
Mike made a longer commitment to the program. Field camps usually
have those long-standing scientists, researchers who organize the field
season and steer the data the crew collects. After the season, they analyze it and publish studies as evidence to better protect the ecosystem
in question. On the islands I'd worked on, the lead researchers didn't
stay all season, but when they were around, they were always full of stories and steeped the crew in history.

For lead researchers, a field camp has also been like a home, an anchor in a turbulent world. I tried to imagine what it would be like to have a connection to a field camp five, ten years down the line, a camp that I'd seen built, grow, and age, and in an ecosystem I'd carefully observed and interrogated for years. Mike's tenure on the island was long, but was in its sunset years—he would soon retire. I wondered if this would be me someday. If I ran a camp and poured my heart and soul into it, living half in more populated places and half on some godforsaken rock, ever shifting between the two, ever reconciling between them. Always disappearing from one life to attend to another. I tried to imagine the moment Mike would get into the Zodiac and leave for the last time. I wondered if there would always be a piece of him lodged onshore with the ice and the snow.

Livingston's human history spans far beyond the camp at Cape Shirreff and the pictures of past field techs nailed on the skua shack wall. The South Shetlands, and Livingston specifically, were some of the first areas of Antarctica to be visited by people.

Common narratives of antarctic exploration center on Europeans as the first people to arrive at the southern continent's rocky shores, but this is up for debate. There is evidence that the sophisticated navigation systems developed by Polynesian cultures that carried them across the entire Pacific Ocean also enabled encounters with Antarctica a thousand years before Europeans laid eyes on the continent. What has been known for centuries by Māori communities and storytellers was reported in a 2021 study that examined "gray literature," evidence outside the usual academic sphere. The study looked at Māori oral histories and carvings to understand the extent of early encounters with the continent. This evidence includes stories of the voyager Hui Te Rangiora and his crew sailing to Antarctica on the vessel *Te Ivi o Atea* in the seventh century. Upon returning, Hui Te Rangiora described the frozen ocean he encountered as *Te tai-uka-a-pia*, in which *tai* is "sea," *uka* is "ice," and *a-pia* means "in the manner of arrowroot," which when scraped looks like snow.

Navigating the ferocious southern circumpolar current posed a significant sailing challenge for early European navigators attempting to "discover" the vast (and, to them, purely theoretical) southern continent they called Terra Australis Incognita. In 1773, the accomplished English navigator James Cook was tasked with finding out for good whether there was any land in the far south. Captain Cook had proved his mettle in a Pacific voyage that lasted from 1768 to 1771, in which he "discovered" New Zealand, claimed it for Britain, and then encountered the eastern coast of Australia, claiming it, also, for Britain. The mandate to push south was spurred by the need to discover additional, exploitable resources to add to the British Empire's hoarded store.

In his quest to find Terra Australis and (you guessed it) claim it for Britain, Cook is credited as the first explorer to cross the antarctic circle, in a voyage that circumnavigated the continent in 1773. While he encountered a few new subantarctic islands, pack ice and weather prevented him from traveling far enough south to lay eyes on the actual continent of Antarctica. Upon his return to Britain, he concluded that the mythical southern continent must not exist. Most authorities took him at his word—such was his reputation—and further attempts to find Antarctica languished for almost fifty years.

In 1819, Czar Alexander I authorized two ships for an exploration of the Southern Ocean, including the search for a continent, and chose Fabian Gottlieb von Bellingshausen to lead the expedition. Bellingshausen was a Baltic German naval officer serving in the Imperial Russian Navy and an avid student of Cook's journeys and achievements. Over the next two years, Bellingshausen also circumnavigated Antarctica, and he lives on in European history books as the first to lay eyes on the continent.

At the time, Bellingshausen was not the only one making the voyage. Also in 1819, British merchant and mariner William Smith had glimpsed

Livingston Island, at the northern tip of the Antarctic Peninsula, when he was pushed south by a storm on his way around the southern end of South America. He didn't take the island's exact position—understandably, he was probably occupied trying to keep his ship on course to Valparaíso, Chile. Upon his return to Chile, his captain, William Shirreff, the commanding officer for the Royal Navy of the Pacific, was skeptical of his young officer's claim that there was land south of the antarctic circle. Nevertheless, Shirreff decided it warranted another trip and sent out an expedition led by Captain Edward Bransfield, with Smith as his pilot. They sighted land once more in January of 1820, coming upon a small promontory jutting out from Livingston Island. Vindicated, Smith named it Cape Shirreff, after his ambivalent superior, in what I must assume was a triumphant flourish. The group of islands that Livingston Island belongs to was named the South Shetlands, after the Shetland Islands, north of Scotland. Bransfield Strait, the stretch of water between the South Shetlands and the Antarctic Peninsula, was later named for Edward Bransfield.

After the existence of Terra Australis Incognita was confirmed, colonial powers launched a number of expeditions in the early 1800s to explore the Southern Ocean and the continent of Antarctica. The expeditions are largely identified by their commanding officers, who are credited with the expeditions' accomplishments and whose names are enshrined in modern maps of Antarctica: Bellingshausen, Cook, Bransfield, Palmer, Weddell, Ross. The official accounts of these journeys are recorded from the captain's perspective and are often infused with the idea of a collectively felt duty to man or country. Often captains demanded that any sailors that kept records, including personal journals, turn in these materials at the conclusion of the voyage to ensure there would only be one official narrative. But who were these sailors?

In the middle of the eighteenth century, England was expanding its navy to maintain its sprawling empire and avenues of colonial trade. At the same time, the privatization of land, the beginning of industrialization, and the mechanization of agriculture left many working-class people displaced and unemployed. Large numbers of dispossessed wage laborers increasingly ended up in maritime employment to meet the high demand for sailors. The prospect of years at sea in the coldest oceans yet discovered was not particularly alluring—the officers of antarctic expeditions often struggled to gather, and keep, their crews. Officers often had to offer rates up to eight times what sailors typically earned in order to man the ships.

For seafarers, these expeditions were a job rather than some holy quest for man and country. They were tasked with the most labor-intensive and harrowing aspects of the journey and often did not share their captain's lofty ideals. The day-to-day work of a sailor in the Southern Ocean was long, cold, and uncomfortable, often completed by first-time sailors who couldn't find work elsewhere.

Crews navigated frequent violent storms, poor visibility, and near-constant subzero temperatures in leaky wooden ships. Sailors shoveled snow off the deck, pumped water overboard, and beat the frozen rigging bare-handed to transfer the warmth from their bodies to the ropes until they were pliable. Thick layers of ice, formed from the constant sea spray, crystallized on the mast and decks and had to be cleared regularly.

Sailing a ship required the skill and judgment to respond to changing conditions even before orders were given by commanding officers, who would stay dry and warm in the upper decks. The sailors' quarters were belowdecks, cramped, filthy, and always damp. They achieved heroic feats of navigation through extreme conditions by relying on collective labor, often coordinating themselves with songs or chants (misinterpreted by officers as signs of good cheer). In the few sailors' journals that have been recovered, a distinct disgruntlement is evident. Multiple times during Cook's expedition stopovers in Tahiti, the captain ordered his ship to sea abruptly and without warning as he knew that there was a strong movement among the sailors to desert and make do on the tropical island. I can hardly blame them.

British seamen developed methods of collective resistance to protest working conditions and were known as some of the most overtly militant groups of the late eighteenth-century working class. While sailors were not the first group to *strike*, it was such a common form of resistance among sailors that the term, used today to describe labor withdrawal in any industry, is named for the act of taking down (or *striking*) a ship's sails to indicate that the crew was refusing to work until their demands were met. Among the basic rights that sailors demanded were more space on the ship, days off on Sundays (apart from basic manning of the sails), and the provisioning of food. Captains had to pay close attention to "murmurs" among sailors that might greet an expedition decision and often changed routes or plans to appease their crews. Ships were social units with distinct boundaries and often delicate politics. While sailors could use their collective power on the ship to negotiate with their officers, they could do little to improve the physical conditions of working in polar waters.

The lives of Southern Ocean sailors in the nineteenth century were ruled by water's phase changes—cutting ice, melting ice, breaking through ice, clearing snow, pumping water, drying one's body or clothes, navigating the churning ocean. In their battle against the elements, sailors were often pitted against just the one, in all its forms, marking the rhythm of their days.

Much has changed in the journey south—today's vessels are much more waterproof, for one. On my own ship, tossed about in the Drake Passage, I often dragged myself outside to the deck to breathe fresh air and bear witness to the shifting skies—clouds like fish scales, like foliage, like the feathery touch of a paintbrush. Southern albatross had swooped and careered on the wind, trailing the ship, appearing briefly and then disappearing into the mist. Even before I stepped onto Antarctica's barren shores, I thought about all the people that had ended up there, and how, and why, and I tried to fit myself among them and feel the shape of the lines that connected us.

The journey to get to the continent

Personal journey & dev.

3.

Early November

Settling into penguin world, Matt and I got busy setting ourselves up for the breeding season: drawing tables in our field notebooks, taking our food and gear to the skua shack, reviewing monitoring protocols. The colonies bustled with activity as the penguins did the same: finding mates, finding a nesting spot, collecting pebbles, preparing for the months to come. The penguins would lay eggs and incubate them for a month before they hatched. Once hatched, the chicks, covered in fluffy down, grew, fed, and loitered in the nest. Penguin parents would take turns going to sea to forage and bring back a meal for their chick. The chicks would grow quickly until they were old enough to thermoregulate without parental shelter, at which point they'd be left alone in the colonies (crèche). The parents would feed them for a few more weeks until their swimming feathers grew in, then abandon them and go on hunting trips to prepare for their own molt. The chicks would leap into the water and "fledge" while the adults stayed on land, shed their old feathers, grew a new coat, then headed back to sea for a long and dark winter.

A couple of game cameras had been set up in the penguin colonies near the skua shack to watch the landscape while the crew was away for the winter. Matt pried open the frozen plastic casing and grabbed the camera cards. Back at the shack, I inserted the SD (secure digital) cards in the laptop designated for the skua shack and opened the pictures. It was an eerie insight into a dark and silent world. In April, the month after the crew left for the winter, a few gentoo and chinstrap penguins were still on the rocks finishing their molt. The gentoos go through their molt later because their breeding season is longer, and they lingered on the rocks for many weeks. In May and June all the penguins were gone, and the light came in at an increasingly dramatic angle. Snowstorms often deposited snow onto the camera and blocked the view. In July, there was accumulating snow, no penguins, and no living thing that I could see. Ice formed a crust that spread from the shore. According to the temperature probes left over the winter at main camp, the coldest day of the year was August 20, when temperatures reached –15.6˚C. I could barely see anything in the camera during August, it was constantly covered in snow. Temperatures only crept above 0˚C between October and March, and even then, not by much.

The peninsula's warming temperatures are more pronounced in winter, which has a big impact on ice formation. Ice might look lifeless, but it is the axis around which the rest of the ecosystem turns. Juvenile krill hide under the sea ice in the winter, grazing on the algae growing on its underside and hiding from predators in pockets made by its irregular surface. The ice itself is laden with nutrients—as glaciers carve through a landscape, they crush and pulverize rocks in their way. As a result, glacial meltwater is full of sediment and rich in iron and other micronutrients. Once this rich water reaches the ocean, it usually remains near the surface and freezes in the winter. In the spring, as the ice melts and sunlight reaches the surface, massive algal blooms proliferate, nurturing krill populations and other marine microorganisms. Peaks in krill "recruitment"—how many new krill are added to the population through breeding—have been strongly linked with the extent of sea ice the previous winter.

Across the Western Antarctic Peninsula, sea ice has decreased in duration by almost one hundred days and in extent by 47 percent since 1979. Without a layer of sea ice, the ocean is more dynamic, and this precious nutrient-rich meltwater is diffused into deeper layers of water, where there is not enough light for phytoplankton to grow. Since the 1970s, a 12 percent reduction in phytoplankton has been recorded in the region where Livingston Island sits. With less phytoplankton, there is less food available for the krill swarms that nourish one of the planet's largest communities of warm-blooded top predators: seals, penguins, and whales. Less ice also creates a feedback loop: without a layer of ice covering the ocean along the coast, marine waters absorb heat from the sun, exacerbating the warming of the region.

When the pictures from September appeared, the light was dim, diffused, and everything looked frozen. We found some stunning shots of a calm, quiet landscape, scant light pouring down from behind clouds onto undisturbed blankets of snow. Only at the end of September did the ice seem to break up, floating off into the ocean and leaving behind tall snow berms on the beaches. By early October, the penguins were intermittently returning to the colonies, and by the end of October those spots were stained pink with their krill-heavy guano. This was where I found the penguins, waiting for the snow to melt.

While the nests remained empty, we set about resighting: finding all the returning penguins with bands on them. A pointy ridge divided the colonies in two, a natural divider for the two seabird technicians—as with every season, we each took one side. I ended up with most of the chinstraps, while Matt had most of the gentoos. All the colonies in the area were part of the monitoring study.

Most birds are banded on the leg, but penguin bands are narrow metal ovals that sit right by the armpit of their left flipper with a unique number engraved on them. Recording the bands of returning penguins tells us about the survival rate of the adult population. If the banded penguins survive the winter and come back to breed, they will be spotted at the colonies. If they are not seen, they are likely dead. Barring predation, these penguins can live to be fifteen to twenty years old, but we rarely saw birds older than ten.

I started to spend a lot of time around penguins on the fringes of my colonies, looking for bands and watching the birds build their nests. Penguins were on the constant hunt for the perfect pebbles: ideally, larger than an almond (for efficiency) and smaller than a tennis ball (for ease of transportation). If all the pebbles on open ground had been snatched up, penguins would prod muddy puddles with their bills, lowering their beady eyeballs close to the ground, hunched over, flippers slightly raised for balance. If they found a pebble, they'd bend down carefully to pluck it from the mud. Pleased with themselves, pebble firmly pinched in their bill, they hobbled back to their nest with flippers aloft, chest out, to ceremonially present the pebble to their partner, carefully dropping it onto the growing nest. When a penguin returned with a pebble, the pair did a crooning mate call to each other, weaving their heads back and forth and bowing to each other. The receiving penguin would inevitably rearrange the pebble from wherever it was carefully placed. Both chinstraps and gentoos were slow and deliberate in picking up and dropping pebbles. With chinstraps, this made a comical contrast to their otherwise frenzied nature.

Pebbles were a hot commodity in the colonies. Penguins often tried to steal one another's pebbles when the birds on neighboring nests were looking the other way. Fights broke out all the time over choice pebbles. The chinstraps in particular were vicious fighters, being more densely nested and more aggressive in temperament. If a chinstrap decided it wanted to destroy another chinstrap, it clamped onto the offending penguin with a beakful of back feathers and slapped the living shit out of it with its flippers. The slaps came in quick succession, *WHAP-WHAP-WHAP-WHAP*, ringing across the colonies. The penguin on the receiving end of the assault tried to break free or run away, often skittering across the ice with the aggressor still attached, attempting to turn around to be the one dishing out slaps. If a penguin lost its grip on back feathers, the two ended up facing each other, each penguin trying to grab hold of the other, and on they tumbled, careening across the colony, aggravating the other penguins and causing all kinds of chaos, until both felt the dispute was resolved, or they got tired or distracted, or one escaped to the welcoming embrace of the ocean.

As the nests took shape and the birds established breeding partnerships, Matt and I spray-painted a bunch of rocks to mark our five-nest study plots. Each colony would have at least one plot, with the bigger colonies having four or five. For the plots, we chose a random spot in each colony away from the area where the previous year's plot had been and dropped a pink rock down amid the penguin chaos. The five nests closest to the rock would be our study plot. In my notebook, I drew maps of the rock and the five nests adjacent to it, numbering each nest. I also marked the known-age nests I'd found with a blue rock, which I'd be tracking daily along with the plot nests. A known-age bird was banded as a chick, as opposed to most of the penguins, who were banded as adults, therefore we knew exactly how old it was. Daily we visited every colony, looking for new banded birds, and checking plots for any eggs laid in the meticulously constructed nests.

While Matt and I were lost to the northern colonies, Sam and Whitney's season centered around the antarctic fur seal harems. Mike was the seal research lead, and as well as running camp, he helped coordinate seal research in the field. CCAMLR protocols largely focused on penguin monitoring, as they were more abundant across the continent than antarctic fur seals, and Mike had developed many of the antarctic fur seal protocols himself.

Antarctic fur seals better resemble sea lions than what many people associate with a true seal (such as the Weddell seal or southern elephant seal) because they belong to a different evolutionary family (Otariidae) marked by external ears. They are far more mobile on land and can hold their upper body up with their fore-flippers, and they have a furry pelage, while true seals keep warm with layers of blubber. Since we'd been at the Cape, we'd only seen fur seal males, who were securing their territory on the beaches and waiting for the females to emerge from the sea. Once the fur seal females came onshore, Sam, Whitney, and Mike were tasked with monitoring the seal harems every day, in the same way that Matt and I monitored penguin plots. Early season they also set about resighting: they surveyed their study beaches, noting the seals that returned with flipper tags, small plastic tags with numbers punched into the webbing of their fore-flippers.

Early season the antarctic fur seal bulls were in their physical prime, huge hulking slabs of muscle and testosterone. Their enormous incisors, mounted on four hundred pounds of sexual aggression, evolved for the sole purpose of tearing at the flesh of opposing fur seals. Once bulls established territory, they had to stay on their beach and guard it constantly, all four months of the breeding season, without once eating or going back to sea. Later in the season they tired and deflated, but in late October to November, they bristled with vitality, everything in their life programmed for this essential gene-distributing moment, fighting for the right to control one portion of the beach and whatever females would take up residence there in a few weeks. Harems could have anywhere from one to fifteen females, depending on the physical space available on the beach and the territory-maintaining skills of the bull.

The lines of each fur seal's domain were clearly designated. They fake-lunged at one another, exhaling dramatic puffs of visible breath, sometimes making contact and ripping skin from flesh. It was a grue-some scene—all the grit and violence fierce competition necessitates. I saw a few males splayed on the beach with flaps of skin hanging open, swaths of muscle and fat exposed, and one with a wound over a foot across. One bull had half its nose torn off. Most were bleeding from somewhere. Flushed with hormones, they didn't seem to even notice the wounds. When we passed the battlegrounds on the way to the penguin colonies, the bulls paid Matt and I no mind, save for a few half-hearted feints, as we were clearly not competition (all skinny and upright as we were), and we crossed rocky beaches mostly unbothered as they eyed one another with suspicion.

The fur seals at the Cape were once victims of a brutal fur trade that decimated their populations. In 1819, when sealers from England and the United States heard of the sailor and sealer William Smith's descrip-tions of the abundant "rookeries" (geographically distinct populations of seals) in the newly discovered South Shetlands, they wasted no time descending upon these populations. In the early decades of antarctic exploration, from the 1750s to the 1900s, science and commerce were closely interwoven. Many "firsts" were accomplished by sealers looking for more pelts to hunt, recording data on the landscape as they went. By the early 1800s, the sealing industry operated in and was quickly deplet-ing rookeries in subantarctic islands such as South Georgia.

So many sealers were looking to make their fortunes from antarctic seal skins that the years between 1820 and 1823 are often likened to the Alaskan gold rush of the 1890s, or the buffalo hunts of the nineteenth century. Historians estimate that a staggering 1.7 million antarctic fur seals were killed in the region during the sealing era. Antarctic fur seal pelts were shipped to Europe and the United States and made into lux-urious fur coats.

After 1823, seal populations were so reduced that sealing dropped off sharply, and merchants turned to whales. Since the mid-1700s, whal-ing captains from Norway and England had been active in the Arctic, but in the last decades of the nineteenth century, whale populations in the north collapsed due to overfishing. The Southern Ocean provided a potentially lucrative alternative: baleen whales migrated south every year from their winter breeding grounds in the tropics to feast on the abundance of antarctic krill. Steam-driven vessels and steam-powered harpoons, invented in the mid-1800s, became essential in the global pursuit of whales. It is estimated that 1.8 million whales were killed in the Southern Ocean between 1900 and 1970.

Scientists don't know much about what the Southern Ocean's ecosystem was like before whaling reconfigured the food web, but they do have one theory: the krill surplus hypothesis. The mass removal of large populations of whales meant that suddenly a whole lot more krill was around for the remaining antarctic wildlife, newly protected under the Antarctic Treaty, which entered into force in 1961. Antarctic fur seals returned to their historical breeding grounds from small populations that had escaped the onslaught of the sealing years, and their populations began to recover. Penguin populations also increased during the whaling era.

Whaling in the Southern Ocean was largely a free-for-all. That is, until military researcher and sound engineer Frank Watlington accidentally captured whale song underwater while trying to record dynamite explosions. In 1966, he passed the whale recordings to biologist Roger Payne, who was so moved by what he heard that he made it his mission to share the recordings as widely as possible. Payne put together an album of whale song, which was incredibly popular. The ethereal, haunting quality of the sounds stirred a larger cultural movement to understand whales as complex, sentient beings, capable of language and music. Judy Collins used them in her song "Farewell to Tarwathie." A nascent Greenpeace seized on the cultural moment for their first campaign: Save the Whales. In 1986, the International Whaling Commission suspended all commercial whaling, which ended the practice in antarctic waters. Today, whale populations in the Southern Ocean are showing signs of recovery, with humpback whales in particular bouncing back to 93 percent of their pre-exploitation levels.

The ecosystem of the Antarctic Peninsula quickly recovered from decades of direct exploitation. On Livingston Island, fur seal populations rebounded from near annihilation, taking off in the late 1980s and peaking in the early 2000s. Mike lived and worked on Cape Shirreff during these times of abundance. He remembered how the beaches were packed with seals, how the population spread inland and up into the hills. He remembered fur seal pups so emboldened by their numbers that they'd explore without inhibition, even hopping inside main hut. At its peak, the Cape Shirreff fur seal population boasted pup counts of sixty-two hundred.

Ecosystems, made up of a set of complex and closely tied relationships, are resilient, but only to a point. Climate change will be the biggest test of that resilience, with vastly uncertain outcomes. Unlike the sweeping destruction of the fur seal trade or the whaling era, climate change in Antarctica manifests itself largely in subtle year-to-year changes, variations only discernible with decades of data to look back on.

After the early 2000s, antarctic fur seals started to decline, driven largely by climate-driven shifts in their environment and increasing competition for krill from recovering whale populations and the krill fishery. In recent years, the annual pup count at the Cape has not even reached one thousand. Mike has told me that he remembers how in the days of booming seal populations, the fur seals were well-fed and more gregarious—puppies were bolder, more curious, more playful. In later years, he noticed that the animals, as populations declined, were conserving their energy. Play is one of the first things to go when an animal has limited energy.

———

Matt saw the first gentoo penguin egg on November 6 in his colonies, ten days after our arrival. An early starter, the nest far preceded the others, and it took a few days for the rest of the gentoos to follow along. We were just finishing our rock hauling and plot setup, with pencil-drawn plot maps in our monitoring notebooks and empty tables ready to be filled with hatch dates and daily adult attendance data. Since chinstraps were far more abundant (thirty-five hundred nesting pairs versus nine hundred gentoo nesting pairs), the study design required that we monitor one hundred chinstrap nests and fifty gentoo nests.

Matt and I each had a "Repro" (reproduction) notebook to keep track of the breeding statuses of our respective colonies. We used a communal "General Penguin" notebook for surveys we did together across the whole study area: population counts, egg weighing, chick counts, diet samples, and device deployments. Our notebooks were thick, yellow, waterproof, about the size of a paperback, and indescribably precious. The dates and numbers inside encompassed the whole purpose of the field season, the culmination of the labor of dozens of people and hundreds of thousands of dollars of federal funds. If they blew away in a gale or got dropped in the sea, then all of this was for nothing. My data notebook deserved a delicate, sheltered existence, but would endure the opposite: getting smeared with penguin shit, dropped in the snow, soaked, battered, mangled in a full pack, stained with food, and regularly set out to dry by the propane heater. Back in San Diego, all the penguin field notebooks from all the seasons got stashed in a bookcase—a wall of battered yellow spines, exuding still a faint whiff of penguin—kept by the seabird program lead.

Every day that followed the first sighting there were more eggs—oblong orbs, tinged green and blue and speckled brown, appeared in the gentoo nests like smooth, bright treasures. Each penguin pair lays two eggs in a nest, the second four days after the first. Once the female lays her two eggs, she goes out to sea to replenish herself, while her mate takes a turn incubating. The first shift is the longest, as the male sits on the nest, allowing his partner time to restore her energy.

The pebbles parents collected so meticulously cradled the eggs, lifting them above the mud and snow. As they incubated, penguins raked pebbles out of the center of their nests with their feet to make the hollow deeper, piling the pebbles up on the sides with their bills to build better shelter. Some nests were castles—mounds of pebbles so high the penguins struggled to climb them, fortresses of stone. Other nests were almost flat, except for the divot in the middle that was the result of penguin nails making a hollow in the ground rather than building up pebbles on the sides.

It's easy to think that humans are the only animals with broad individual variation, that there are a million ways to be a person but only one way to be a penguin. But as always, the closer I looked, the more detail I discerned, and I found just as much variation in penguin personalities as in people. It was also all too easy to project human qualities onto penguins just by looking at nests: the overachiever, the control freak, the chaotic artistic type, the hot mess. Some young pairs that didn't yet have their timing down popped an egg out before their nest was ready and incubated it on a handful of pebbles hastily gathered from their neighbors (what Matt and I called oopsie eggs). Once the eggs appeared, they were guarded so tightly that we'd rarely see them.

At the beginning of the breeding season, feathers fall off a patch of skin under a penguin's abdomen, right where the eggs will nestle during incubation. This bare patch of skin, called a brood patch, has a higher density of blood vessels to bring more warmth to the eggs. When the birds stand up, the brood patch is covered by other feathers. The birds will regularly rotate the eggs in their nests so that all sides get a turn pressed against the warmth.

While the two penguin species under my purview had similar egg-shielding habits, they differed in their life history. Gentoo penguins are more flexible in the timing of their breeding season than chinstrap penguins and can adjust their egg laying by days or weeks. If it's a low-snow year, they will begin the breeding season as soon as there is bare

— 39 —

humanizing penguins

ground; if there's more snow than average, they'll wait. Gentoo penguins are also more asynchronous, meaning that they don't all lay at the same time, so risk is distributed across the breeding population.

Chinstraps, on the other hand, don't vary much from their pre-programmed lay dates and all lay at around the same time. If it's a big snow year and the ground of the colonies has not melted, they will either skip breeding that year or attempt to breed anyway, building nests on the snow. The eggs may fail if they grow cold and wet from too much direct contact with snow or water. Researchers fear that chinstraps are more vulnerable to the effects of climate change because they will be less able to adjust as weather patterns begin to shift.

Once the egg is laid, the long task of incubating begins. The penguins passed the time looking for more pebbles, but while they sat on the nest, their search radius was vastly reduced to what they could reach with their bill. Sometimes I'd see a penguin with its eye on a pebble that was a little too far away, that had perhaps tumbled from a neighbor's collection, and it'd stretch out its neck to its maximum length, push forward with its toes, and reach with all its might. Sometimes it'd snag it and sometimes it just couldn't, but not for lack of effort.

When a penguin's mate returned to the nest from a foraging trip at sea, the pair crooned and bowed to each other in greeting. The incubating penguin, hungry and disheveled, stood up off the eggs and stepped to the side. The incoming penguin shuffled carefully onto the edge of the nest, feet right up against the shells containing its precious progeny. It bent down and scooted the eggs up onto its feet with its bill, then lifted its chest over the eggs and lay down, covering them with its brood patch and sheltering them in feathers. The other bird hung around for a few minutes, making adjustments to the nest, perhaps collecting that pebble it hadn't been able to reach while incubating, before taking off toward the ocean. I tried to imagine the sweet relief of seawater washing off the grime of the colony and the anticipation of the meal to come.

On my daily rounds I checked all my plots, six gentoo plots and nine chinstrap plots, as well as the 38 known-age nests I'd found in my colonies—113 nests in total. After both eggs in a nest were laid, I'd stay for a few minutes in case a bird lifted itself to stretch and I got to see the nest contents. Once every four days I lifted tails on the nests I hadn't seen—holding the tail bristles and raising the penguin's butt just enough to get a clear peek into the nest. Four of my plots were "nondisturbance," which were to be monitored without any tail lifting. These were a test of patience—I stood a few meters away, at the edge of the colony, and observed. Some penguins were stoic like statues, some fidgeted, some eyed me warily. It was like a game in which I had to out-wait their patience, waiting for a penguin to move enough for me to see what it was hiding under its feathers.

I loved watching the penguins, and shrouded nobly in my scientific purpose, I lingered near the colonies after my plots were checked, feeding my heart. What made me fall in love with fieldwork in the first place was how, in the absence of the screens and noise of contemporary society, my world snapped into vivid, immediate focus.

My first taste of fieldwork was a summer internship with the Fish and Wildlife Service in the northern woods of Maine, surveying streams for a salmon habitat–restoration project. I applied because I wanted to work outside, and I'd read a book about Maine as a kid that made it sound like a sea of trees thick with animals (I was not disappointed). Toward the end of the summer, I stayed on a seabird-nesting island off the coast called Eastern Egg Rock. It was tiny—you could walk across it in five minutes. People slept in tents on platforms on the rocks, with a central cabin for eating and cooking and data and hanging out. We were monitoring the reproductive success and diets of reintroduced puffins, as well as keeping a general account of the other species that alighted on our remote rock. Life on the island was simple. Vegetables were stored under the deck, dishes washed with seawater. The roof of the cabin was our lounge. Sunsets on the water exploded from the sky. I couldn't believe that this was something people actually did for a living.

As on Cape Shirreff, human history was a core part of the ecosystem's history on Eastern Egg Rock. Hunters had wiped out puffins
from Egg Rock and nearby islands in the late 1800s. Puffin breeding
populations had only been brought back through the long and painstaking efforts of Dr. Stephen W. Kress, who, beginning in 1973, planted
puffin decoys and played recordings of puffin sounds to lure in unsuspecting puffins from the two remaining islands that still hosted breeding puffin populations. The Audubon's Project Puffin has brought back
a thousand puffin pairs to three islands off the coast of Maine, where
they enjoy their homelands in peace—except for eager field technicians
who constantly watch them through binoculars.

In the afternoons the monitoring crew and I would disperse into the
blinds that dotted the island and watch for puffins to return from the
sea with loads of fish in their bills. I rushed to write down the number
and species of fish before the puffins disappeared into their burrows
among gray boulders. We were studying their diets, as well as monitoring their reproductive success by keeping track of the burrows and
the chicks that eventually emerged from them. The program also recorded data on the rest of the species that lived there, as part of a larger
ecosystem-monitoring effort. Tern nests covered every inch of the island, and laughing gulls stood in groups on boulders near the water. At
night, a storm petrel would arrive and shuffle into its tunnel under my
tent platform, purring at its chick. The soil, fertilized by all the nitro-
gen deposited by seabird droppings, exploded with grasses and flowers.
The ocean gave life to land in an ancient, well-known dance. I loved the
way the birds anchored me to the island, the way we made our home
from the same stones, and the way they connected me to the ocean and
everything beyond this tiny rock.

In the mornings the crew and I headed to a short fence enclosing a
group of nests and weighed tern chicks to check on their growth. Consistent mass measurements can indicate how successful the adult terns
are at collecting enough food for their chick and therefore tell us about

the abundance of their prey (small fish, crustaceans, squid, inverte-brates) in the ocean.

The first time I held a tiny chick in my palm, I could feel its heart-beat through its tissue-thin skin. It was as big as a golf ball, all fuzzy down, barely able to lift its head, its wings soft nubs and its eyes barely open. The life in this tiny thing seemed so tenuous, like a tender, breath-ing seed. I could see a small white egg hook stuck to its minuscule bill—even though it seemed helpless, it had already fought its way out of its egg, a task far greater than anything human newborns have to under-take. The impression of that chick stayed with me for years, embodying so many things stubborn and precious, but easily broken.

My world on Antarctica was much larger than Eastern Egg Rock, en-compassing a stretch of land that took an hour to cross, from the glacier half a mile south of camp to the penguin colonies a mile from camp on the north shore. The borders of my world were geographically distinct. I grew accustomed to the hulking shape of the tabular iceberg floating to the west and to the white mountain range that rose from the ice-covered mainland, nineteen miles away. Early in the southern summer, the sun beamed down unobstructed, and the first day we glimpsed the moun-tains, Matt and Whitney rushed out with their cameras to capture the view. It was evening, and the light was gentle, bathing the snowy west-ern flanks of the mountains in pink. Whitney and Matt peered at the blue skies skeptically, always bracing for the storm they promised us was imminent. The previous year the sight of the mountains had been rare.

While our own landscape was quiet, a storm brewed elsewhere. It was election season. Hillary Clinton and Donald Trump were compet-ing for the presidency. For news, and as an exception to our data-driven rules, Palmer Station emailed us a PDF of the *New York Times* daily di-gest every day while we were in camp. Normally we were not to send or receive any attachments, only plain-text emails. But a few pages of daily news was considered a tether to the world worth maintaining. On the night of the election, like most of the communities we came from, we thought we would get to celebrate our first female president. We didn't have CNN to watch or the internet to refresh, so we took turns calling people on the satellite phone for updates. Late at night it was my turn and I dialed my mother, who was living in New York City with my father (their latest posting). She was in tears. She told me Trump was winning all these states he wasn't supposed to win, how it was looking bad, and by the time I hung up a brick of anxiety was in my stomach. I came back in and reported what I knew to the crew. We stared at each other in disbelief, thinking, This cannot be happening, this cannot be happening, and went to bed without knowing for sure what the next four years held in store. We were an hour ahead from New York, and while the ballots were counted and results announced late into the night, we slept in the hut on a distant island, wind howling outside.

In the morning, Mike called his family on the satellite phone and our worst suspicions were confirmed. We sat in stunned, stony silence over breakfast. It may sound redundant to say that we all felt extremely disconnected and alarmed, since we had chosen the isolation of Antarctica, but we all came from somewhere and had to go back to somewhere. The crew was all from the United States, and this was not that somewhere that we thought we knew. It was strange to be so cut off from the world in such a historic moment. The island was calm, but I was afraid.

My concerns for the new presidency spanned far beyond the fate of the Antarctic Peninsula, but after a month in the field, I couldn't help but fret over what this would mean for the penguins. Field-workers are intimately acquainted with a world few people know, and we see closely the toll that our society has on the species we study. It's impossible, in the field, to think we are separated from "nature," that what happens in human societies has little bearing on the ecosystems that make up our world. The isolation from society I might feel working on such a remote island is an illusion created by a solitary day-to-day existence—our societies are deeply intertwined with what happens on the continent, and this is never more evident than in the massive impact of climate change. Deep in an ecosystem that was slowly unraveling, shifting and churning, I often felt unmoored and overwhelmed, especially when the forces driving these changes seemed so big and so far away. The election only heightened my sense of urgency.

4.

Mid-November

Wind was the first thing I heard in the morning, along with a door opening and closing as Whitney got up first and went out to use the outhouse. Sounds reached into my awareness through the fog of sleep. Then: the lighter button of the propane heater pressed, a metallic clang sounding at least twice until it caught. I heard the kettle being lit and muted footsteps on plywood. Whitney was brewing coffee. The old, damp smell of socks and mold faded into the earthy scent of coffee.

My socks hung drying on a line over my bed, and my long underwear draped from the rod of my bunk curtain. I picked the pair of socks that were the least stiff and pulled on the long underwear. Sam and I stumbled from our bunks at around the same time, usually a bit past eight. Matt would stand by the door, wiping the condensation from the window with a squeegee and peering outside. He ate two packets of instant oatmeal every morning without fail, washed down with a steaming mug of coffee. He was not a real person in the morning. I did not attempt interaction. I already knew this about him, but the rest of the crew learned it quickly. Sam once tried to have a morning conversation with him and, after a terse and dismissive reception, left thinking that Matt hated his guts.

The one thing that everyone did soon after emerging from his or her bunk was to check the weather display. The weather dictated the fluctuations of our lives, as we would be outside for most of the day. Any lingering doubts were cleared up by the morning dash to the outhouse, hands balled up inside a sweatshirt, bracing against the wind and squinting at the light. If I wasn't fully awake before that, then I sure as hell was afterward. Outside, a troop of penguins would be walking by camp or skuas would be careening acrobatically over the beach, looking for carcasses. Sometimes it was snowing, sometimes it was foggy, sometimes ice would pelt my face. Almost always, it was windy.

Taking turns to make breakfast, we wove past each other like interlocking links in a chain. If there were no chores around camp, Matt and I would suit up for our commute to the penguin colonies. I grabbed the radio I carried everywhere from its overnight charger and refreshed the snack stores in my pack. Hiking out, I always had to be slightly cold because soon I would be sweating from the walk: gradually uphill to a ridge, down to Chungungo Beach, up another ridge, between rolling hills, over a last ridge, and to the skua shack. Once there, I changed into my penguin rubbers, overalls, and a jacket made of stiff waterproof rubber, along with dedicated penguin-only boots. We had a quick mug of tea if there was time and went out to attend to the penguins.

Once a nest was confirmed active, when eggs appeared, I banded one bird at each of the seventy-five plot nests I was tracking so I could discern between individuals. It was easiest to do this when the penguins were incubating because they didn't run away. At the beginning of the banding period, Matt took me out to his colonies so he could teach me how to handle and band the birds. As gentoos had started laying first, we started with them.

When we'd worked together on St. George Island in Alaska, we'd catch a bird with a long pole with a noose at the end, maneuvering it tactically while dangling precariously at the edge of a cliff, looking down at the target kittiwakes, a species of small coastal gulls. For the larger gulls, we stood at the base of a cliff with another large pole, trying to control the wobbly tip, often while standing on a ladder. For other birds, we set traps: the least auklets, small, mottled seabirds the size of a robin, were captured by a tangle of nets tied to rocks with nooses that tightened around their ankles when they walked over them. For petrels we strung up a thin net between trees, carefully disentangling the birds' delicate limbs when they flew into it. In my field experience to date, catching birds had always been an elaborate operation, requiring gear, preparation, and time.

Matt and I walked up to a gentoo colony, and I watched as he bent down and simply plucked a penguin from its nest and tucked it between his legs. It was a calm gentoo, and Matt had the band on its flipper in less than thirty seconds. He plunked the penguin down and stepped off to give it space to gather itself and return to its eggs. It was the simplest bird capture I'd ever witnessed.

"You can just . . . pick them up?" I was amazed.

During some supervised penguin banding I learned the right spot between my thighs to pincer the penguin and the right angles to bend the metal of the band so its edges were tight against each other. Satisfied that I was ready, Matt released me to band in my colonies. When the metal band was closed and secured, I plunked the penguins down again by their nest, and most settled back onto their eggs. Once, I put a chinstrap down after banding it, and instead of scurrying away, it stood there slapping my leg viciously in bitter revenge.

I found that penguins differed in their reactions to being hoisted between human legs. Some were calm, mildly befuddled at how they got a foot off the ground. Others acted as if they were possessed, squirming and slapping and biting. Penguins are beefy birds, sleek bullets of swimming muscle, torpedoes of power, and they slapped impressively hard. Their slap was powered by the same muscles in their chest that propel them through the water. A sharp stinging whack by a penguin flipper in the bitter cold could temporarily inactivate a hand.

Once all my birds were banded, my daily rounds consisted of walking around the periphery of my colonies, always in the same order; reading bands; peering at growing nests; checking my plots; noting which penguin in a pair was on the nest, banded or unbanded; and looking in the crowd for banded birds I hadn't already recorded. I spent hours on my own with the penguins, weaving through their colonies, close enough to touch them, but far enough not to disturb them. I did not miss human companionship—I was happy amid their ecstatic energy.

My last colony was on the high ridge, which I climbed along the penguin trails, both me and the chinstraps huffing as we worked our way up the hill. On beautiful days I ended my work with the stunning view of the peninsula below me, all gently rolling hills, the distant square shapes of camp perched by the ocean, the glacier beyond it, and Livingston's massive snowy mountains in the distance. Everywhere else was ocean—in the expanse of dark blue, swells kicked up by the wind looked small from a distance, roiling and churning. Sometimes I caught sight of a whale fluke as a humpback hunted for krill. On stormier days, which were more common, the view from the most exposed part of my rounds was simply a wall of snow or fog.

After Matt and I had each done our separate rounds, we sipped tea in our office chairs back at the skua shack, looking out at the colonies and having long, meandering talks—about the books we were reading, our dreams for the future, my literal dreams, our families, food, stories from our other seasons. Our friendship was as comfortable as the wheelie office chair I collapsed into, holding me easily and comfortably just as I was.

The shack sat midway down a slope that gradually leveled, leading past the colonies to the beach, where the male fur seals were still locked in battle. About once a day, Matt peered out the door, thumbs in the straps of his bibs, and said, "I think we need to go check the snow." I grabbed my own jacket and was out the door.

We slogged up the hill with a small red plastic sled. I sat in front, Matt dropped in behind me, and when we were ready, we lifted our anchoring feet and flew down the hill. It was such a thrill that I couldn't help but cackle madly throughout the ride. When it had been a cold night and the snow was icy and hard, we went far; when it was wet and slushy, we barely made it past the shack; and on off days we might veer to one side and tumble over each other in a mess of gloves and legs and clumps of snow. If we reached a small flat bit, we'd try to keep the sled going by paddling with our hands, calling this our avalanche-prevention study.

The snow didn't last long—by mid-November, the snow covering the rocky hilltops of the gentoo colonies had already almost entirely melted. In its place was the characteristic substrate of a penguin colony, which had many nicknames. I called it penguin butter and Matt called it penguin pudding, and we sat in the skua shack and argued about which word best represented the muck we tromped through every day. It resulted from decades of penguin poop, mud, and constant precipitation. It was thick and stuck to everything. The smell was strong, like fermenting shrimp and manure, fishy with an avian touch. At first, it was overwhelming, but over the season I got so used to it I no longer noticed it at all.

The program had long lost any initiative to try to clean and reissue penguin clothes for the next season. The stench became so impregnated in the fibers of anything it touched that cleaning it off was simply not possible. Our penguin gear would get incinerated on the ship when we closed camp. Matt and I made vain attempts to wipe off the grime with snow or rinse it off with salt water, but it was only for our own sake in the months to come. There was no washing machine here. Our washing infrastructure consisted of two buckets and a washboard, which I came to learn was an actual board with ridges against which to knead the grime off clothes and not just a way to describe someone's abs.

The seal crew, as per tradition, was disgusted by penguin pudding, while we got covered in it every day. There was no other way to do the work. Time-honored camp protocol demanded that penguin muck had to be kept out of main camp at all costs. When we got to the skua shack, we had a whole set of rubber outerwear—boots, flannel, jackets, and hats—to be worn in the penguin colonies only. Our days were marked by the morning change into penguin clothes and the afternoon change out of penguin clothes. "Ugh," Matt would say, wiping his hands on his bibs after taking off his grimy boots, "gross." The plywood skua shack floor was already seasoned with over two decades of penguin muck. The floor was always wet, or at least damp. Any food that we dropped on it was instantly dead to us. It stripped the color off an M&M within an hour. The floor digested. It was alive.

The five of us settled into camp and into one another. Whitney, short and red-headed, was ever cheerful, always there with a smile and cracking dark jokes as if it were her job. Sam asked probing questions about everyone else's life and shot indoor hoops with an ancient foam ball. Sam and I often caught up with each other in the evenings while shoveling snow outside. We were sharing encountering all of this for the first time—we laughed at the absurdities of our job and Matt and Whitney's eccentricities. We'd often share tidbits of what we'd heard from them on how to keep camp running, or ideas for the next season, prefacing statements with "Okay, so next year we have to remember to . . ." Sam was always generous with his energy, a good listener with an easygoing, grounded presence that made me glad we'd be working together the following year.

Matt was ever quiet and thoughtful, while I simmered with giddy energy, trying to get a handle on everything. Mike often receded into his inner world to the exclusion of all else. I could call his name from only a yard or two away and he would not notice. Sometimes he'd want to share something with us, and he'd look up as if he were seeing us for the first time, as if he'd been underwater and just broken the surface for a lungful of air. He'd burst out with the classic "Hey! You guys!" that usually prefaced comments or announcements, and we'd pause and stand to attention.

In the first few weeks, Mike could often be found sprawled on the floor with his head deep in the nonfunctional oven, tools littered around him, door unscrewed, troubleshooting, trying to fix the appliance from an old, faded manual and from text emails from coworkers who could look something up for him on the internet. He was at it for days, but there were no breakthroughs. We all tried our hand at the problem, and while we didn't succeed, I learned a lot about the anatomy of an oven. Mike was both a brilliant cook and deeply invested in his sourdough project, and a whole season without an oven was simply unfathomable.

Instead, we started using the grill as an oven (what we dubbed the groven) and experimented with various settings and pan configurations to attempt an even bake. We moved the grill into the stay-wet room to shelter it from the wind, which had the added benefit of warming up the room for anyone who was showering. This efficiency came with limitations—the shower could only last as long as it would take to bake whatever the designated cook had put into the groven, lest our precious food get burned.

I averaged a shower every two weeks. Sam's and Whitney's interval was a little shorter, Matt's a few days longer. If my intention was to be clean by nightfall, I had to plan for it all day, coming home early, melting snow, heating the water on the stove, hauling it out to the stay-wet room (preheated with a propane heater if the groven was not in use), and finally standing under the dribble that the small pump emitted as the water level in my bucket slowly dropped. Decent hygiene seemed impossible when even the basics were such a process.

In the evenings we took turns cooking dinner and gathered around the foldout table in our plastic chairs, wolfing down whatever the day's cook had concocted from the cans in the pantry, the produce in the freshies room, and the supplies in the chest freezer.

The sun dipped low late in the day but didn't sink below the horizon, just getting deeper in color and making more dramatic, angled shadows. I slept in light and woke in light. Outside the iceberg turned as the wind changed, showing off its many faces to the lingering sun. When large chunks of it broke off and washed up, we chipped some off and used it to chill whiskey.

After dinner, we lingered at the table, sipping wine, playing cards, reading, chatting. Home settled around us. Someone would be tapping away at the dedicated "email computer," the only laptop hooked into the satellite email service, and someone else would be standing at the back door, staring out the window. The Christmas lights would be on, music would be playing. It felt quotidian and normal, as if it could be happening anywhere.

Sometimes we dug for a movie on the camp's hard drives and played it on a small projector against a screen that rolled down. We set up camping chairs and grabbed our flannel blankets and curled up with a hot mug of tea for the show.

Outside, on the beach, the fur seals would tuck in for the night. A big wave would move toward the beach, a wall of water rising, and through its translucence I could see strands of seaweed and seals, who'd ride it with heads facing the shore, moving as one with the water.

Sometimes I wondered if I was just there for the lifestyle—did I just like the way my days unfolded on a remote island, where my purpose was clear, where my life was simple, where I was directly involved with the resources that sustained me, where my connection to an ecosystem was palpable every day? Did I simply want to feel close with other species, or was I interested in the questions that the data was designed to answer: How is climate change impacting this remote species? How are the populations of seals and penguins changing? What are the vulnerabilities that emerge from their life history? Where do they forage and what is the quality of those foraging grounds?

The science was critical, but it sometimes felt distant and formless—in the field, you had to love the job itself because the job defined the texture of your life. Science was my excuse to slide down a hillside that ended in penguins. To adjust my lifestyle in the kinds of dramatic ways that globalization has diluted, in the kinds of ways that respond to a distinct geography, to a specific landscape. As a kid I moved through so many big cities that they all started to feel the same—busy streets, multistory buildings, the bustle of people, subway trains, the same transnational brands occupying city corners. I felt the resident ecology muffled under layers of concrete. A sense of place rooted in other species felt disparate and elusive. I was ever peeking at the weeds that grew from the cracks in the sidewalk, trying to catch a glimpse of the birds that flitted around in urban trees, growing cherry tomatoes in containers on cement patios.

When I started working in the field, I felt nested within a web of other species, grounded in place and biome in a way I'd always wanted to be. I longed for far-flung outposts in what I thought of as wilderness, where human impacts were absent or minimal. Hailing from sprawling capital cities, I, like many environmentalists, associated a human presence only with ecosystem destruction and degradation.

But wilderness as I understood it, rather than some kind of pure state of nature, is a complicated concept steeped in human history. "Wilderness" was glorified in the Romantic age (1800–1850) by Europeans as a counterpoint to growing industrialization. Romantic-age writers wrote of wilderness as a sublime landscape where one could encounter God. In a now-classic 1996 essay titled "The Trouble with Wilderness; or, Getting Back to the Wrong Nature," William Cronon describes how these Romantic-age ideas of wilderness were carried to North America and combined with colonial narratives of the "unsettled frontier"—a proving ground for true manhood and the only place one could be free. Transcendentalist writers in the 1820s and 1830s were inspired by Romantic poets such as William Wordsworth and elevated North American wilderness, a supposedly pure and untouched landscape, as sacred.

When I first started working on remote islands, I loved reading transcendentalist authors. I resonated with the poetry Thoreau and Emerson evoked from the earth. I carried *Walden* around with me for *years*, an old copy battered from being tucked into so many field packs.

I'd first read *Walden* in college and didn't think much of it; it seemed like just some dude in a forest ragging on society and growing beans. But I took it with me to Alaska, just in case. My first full field season, on St. Lazaria, a temperate rain-forest island where storm petrels came in at night and nested in burrows underground, I encountered a vibrant world with which to map Thoreau's thoughts. The simple one-room cabin in which our crew lived and the forest around us echoed his world much more closely than my dorm room in a sleepy suburban college town. In the field, the crew and I sipped tea from Sitka spruce needles and snacked on the salmonberries that exploded into our trail. I could hear the words "Talk of heaven! Ye disgrace earth."

It seems like a caricature now: an upper-middle-class white student in a liberal arts college reads *Walden*, stares at trees, and feels spiritual. But there is still something about Thoreau that moves me: "You find thus in the very sands an anticipation of the vegetable leaf. No wonder that the earth expresses itself outwardly in leaves, it so labors with the idea inwardly. The atoms have already learned this law, and are pregnant with it."

The popularization of wilderness supported the idea that the North American landscape was pure and unmodified, thriving without human influence. But this could not be further from the truth: indigenous cultures across America practiced diverse and extensive landscape management, including shaping grasslands and the range of bison herds, using fire to create open woodlands, and maintaining vast food forests. Yet, conservationists such as John Muir and politicians such as Theodore Roosevelt weaponized the idea of wilderness against Native peoples, forcibly evicting them from many of their cultural homelands to establish national parks: in two instances, Ahwahnechee communities were forced out of the Yosemite Valley in 1890 and Black-feet communities were forced out of Montana's Rocky Mountains, in what would become Glacier National Park, in 1910. Conservationists then, as now, glorified landscapes without people because they framed the destruction that industrialization, capitalism, and colonialism caused to ecosystems as the impact of "humankind" as a whole instead of the impact of one particular cultural paradigm. Cronon writes:

> Only people whose relation to the land was already alienated could hold up wilderness as a model for human life in nature, for the romantic ideology of wilderness leaves precisely no-where for human beings to actually make their living from the land. This, then, is the central paradox: wilderness embodies a dualistic vision in which the human is entirely outside the natu-ral. If we allow ourselves to believe that nature, to be true, must also be wild, then our very presence in nature represents its fall.

Before a field season, I also tended to pick up books of stories, his-tories, and cosmologies from the Indigenous cultures that developed in the regions I was working in (Tlingit, Inupiaq, Aleut, Hawaiian).

Thoreau's pretty words weren't the only narratives swirling around in my head on St. Lazaria. In the Tlingit stories I was reading, *kushtaka*, shape-shifters, moved between human and otter forms, trying to cap-ture the souls of dying people. When a storm was coming, the otters would gather on the lee of the island and tangle themselves in beds of kelp, and we'd know to bring things inside. In the calm before the storm, we'd hear the otters' tap-tap-tapping as they opened shells with rocks. I thought of how eerily human their gestures were, fussing with shells on their bellies, gathering up their cubs for the night. I tried to imagine the texture of Tlingit life before colonization as the crew and I fished for rock cod and made fish prints on the door of the wooden hut where we lived.

When I worked on Midway Atoll, I became close with a friend and crew member who had native-Hawaiian heritage. She told me how Midway and the surrounding islands were known as the home of the ancestors of Hawaiian culture. She told me about Polynesian voyages on double-hulled canoes and how skilled navigators found their way around the vast Pacific Ocean by orienting themselves with stars, currents, seabirds, and wind. We learned the Hawaiian names for all the plants we worked with and stared up at the night sky, trying to find constellations and imagine them as a map.

Transcendentalists understood nature as an entity to be respected and worshipped, but one that existed outside human society. In Aleut, Tlingit, or Hawaiian stories, I never encountered even the concept of nature—a category for every living thing on the earth except people. The division between man and nature simply did not exist. Reading Indigenous mythology and trying to understand Indigenous worldviews helped me interrogate this false binary. But Indigenous books and ways of knowing, so different from my own, were a testament to the survival of Indigenous communities. They could not be separated from the violent history of my own ancestors—settlers and colonizers tried their best to eliminate the Indigenous peoples, cultures, and ancestral knowledge that now gave me such insight into my own mind and into the places in which I worked.

Antarctica is often referred to as the last great wilderness, a continent where the near absence of humans elevates it to a sacred status. There have been visitors, but no culture has developed in Antarctica, no language has flourished to describe it, no person has been raised there and acquired the rootedness bestowed by local ancestors. *Antarctica* means "opposite to north," and the continent has served in the popular imagination as a counterpoint to all things human and organic. Lauded as a symbol of purity, remoteness, cold, and extremity, it also has an intractable novelty: all who have ever worked or visited there remember the first time they stepped onto the continent. The vast expanses of ice and windswept hills may feel foreign, but they are far from alien. They are as much a part of our living earth as the equatorial rain forest or a temperate grassland.

Making my home in Antarctica sometimes felt like a contradiction— what does it mean to live a domestic existence in what is often called the planet's last or ultimate wilderness? It was surreal to wash dishes in the kitchen sink while looking at a line of penguins walking by, to call antarctic fur seals my neighbors, to fall asleep to the roar of circumpolar wind. Before it became habitual, I reeled to make sense of it.

Eventually my cold and distant home didn't seem complicated at all—humans are remarkably adaptable and soon I settled happily into the close coexistence of the island. On one hike back from the colonies in November, I saw whales blowing spouts in the waters past camp. They must have been just over a half mile offshore, at least four or five groups of them, feeding. Gulls hovered and swarmed over each group, hoping to snatch whatever food the whales brought to the surface. Two or three enormous mouths would lunge out of the water at exactly the same time, and the upper jaw, black and shiny, would seal the lower jaw, bulging and streaked in white. I climbed a rocky spire and watched them feed. A long, flat iceberg was some one to two miles from camp, covered in penguins. The whales were feeding on all sides of it, spouts shooting out of the water. The clouds were thick and low, sealing the horizon not far beyond the whales, making this all seem close, contained, intimate, as if we were huddled together under a fluffy gray blanket: the gulls, the whales, the penguins, and I.

HATCH

5.

Late November

My days became like foamy waves lapping onto a dark rocky beach: steady, rhythmic, and dissipating quickly. I woke to a windy, raw white world, monitored my penguins, hiked back to camp, stuffed my body with fuel, chatted, slept. I became thicker, my body packing on the fat to better keep me warm. My legs grew strong from all the hiking and hauling, my shoulders taut. While before I huffed and panted, soon I mounted the hills mechanically, pumping away, eager to get to the top.

I charted wind speeds to the way my body moved across the landscape. I could barely feel anything below twenty miles per hour. The breeze was brisk and gentle until it reached twenty-four to thirty miles per hour, and by thirty-five it was strong and forceful, a moderate gale. At forty-five we were square in gale territory, a force that would've broken twigs off trees if there were any. At fifty I staggered forward, my body bent over double to lower my center of gravity and be closer to the ground. Gusts of sixty, classified as a "whole gale," were vicious and unforgiving, making it hard to move at all. Anything above seventy-three was a straight-up hurricane. Matt told me readings at the skua shack reached ninety miles per hour during a storm the previous season, a figure so high it was hard for me to even imagine the force of the wind and what it might do to a person alone on broad, empty hills.

On windy days Matt and I, weather-beaten, sat in the skua shack nursing ever-present mugs of tea and listened to the wind whistling through the cracks in the plywood walls. We watched the anemometer's reading on our weather display panel. Vapor from our tea mugs and shed socks rose into the musty air. A gust—the floor shuddered, the lid to the attic popped up for a second—and Matt grinned: "Ooh, fifty-two mph! That was a good one. Come on, sixty!"

On the very occasional windless day, I bathed in the silence. It felt like the pause between an inhalation and an exhalation, like the empty space between words, like stillness between wingbeats. Sometimes a thick fog settled over the Cape, creating a whiteout. White snow was below me and white fog around me, lit by diffused white light, no shadows and no source. Without wind all I could hear was my breath and my footsteps on snow. With no context and no reference points, I could be anywhere and anything. I wondered if this was what it was like to die, or where you go afterward. In pure light I was just one more creature tracing a trail across blank white hillsides.

By late November, the snow cover was quickly disappearing. Every year all the snow melts from the ice-free land during the summer months in the South Shetland Islands, part of the annual cycle of water, snow, and ice. The sun shone on spots of bare earth, warming up dark rocks and soil, which absorbed the light, warming, warming, so that the edges of snow that bordered bare earth began to warm also and melt and recede. The melted snow streamed in rivers down the hills, snaking through dark ridges on its way to beaches and the sea.

The soil beneath the melting snow was dark brown, almost black, dense and heavy with water. Green-blue, thread-thin lichen reached upward from tufts tucked between rocks, and carpets of soft moss—

yellow, rust, green—hugged water-saturated soil in scattered patches. The land revealed itself as if it were shrugging off a fluffy white jacket.

From the hilltops to the rocky shores, the short antarctic breeding season roiled onward. In late November, the antarctic fur seal females began to arrive on the beaches. Compared to the males, with their hulking form, the females appeared soft, elegant, and iridescent, their shorter, finer pelage glistening in the sun. Even to my human sensibilities, they seemed sleek and sexy. Fur seal females gave birth within the first few days of arriving onshore, peppering the beaches with tiny squirming proto-seals.

On the beach, the melting of the snow exposed enormous whale bones scattered on the rocks, left over from the whaling era. A huge jawbone the length of several people lay near my penguin colonies, concave side facing the ocean. Sometimes little fur seal pups curled up inside the porous hammock of ancient bone and fell asleep there, whiskers twitching.

Through the end of November, most chinstraps and gentoos were totally dedicated to their eggs: they sat through blizzards, getting covered in snow and frozen in place; they sat through gales, back feathers blown every which way, head facing away from the wind; they sat through sunny days, heat baking their black coats, bills open, panting in an effort to cool off. For a month, they sat, switched, hunted, switched, and sat. Inside the egg, safe and warm, a fetus was growing. Eyeballs, then a small, hard bill, tiny flippers, two wrinkled, webbed feet, the beginnings of downy feathers pockmarking delicate skin. The pre-penguin curled in the shape of an egg, even before it was constrained by it, floating in amniotic fluids. The penguins and I waited while these tiny creatures developed, attending to the shells that housed them, watching for the telltale star-shaped crack.

Besides penguins, we also kept tabs on the skuas, large predatory seabirds that take up residence at the Cape in the summer. The face of a skua is somewhere between that of a gull and a hawk, black eyes ever scanning for vulnerable penguins to pounce on. Skuas are a mottled brown, with a range of light and dark feathering, but all generally the size (and roundness) of a football. They have curved beaks made for tearing flesh and a withering, pointed gaze. They were the island's main scavengers and nested and bred at the same time as the penguins, gorging on the eggs, chicks, and carcasses of other species. We often saw skuas flying over our peninsula, careening precariously as they navigated forceful gusts, kings and queens of the wind.

Each skua pair had their habitual hunting grounds, and many lingered in the penguin colonies, on the prowl for eggs. To steal an egg, their hunting tactics required coordination between a pair. Two skuas would land near a nest at the edge of a penguin colony, stressing the incubating penguin, who would snap, caw, and try to lunge (without leaving the nest) at whichever skua was closest. After hovering just out of reach, the skua behind the penguin would step forward, grab hold of the penguin's tail feathers, and give a little tug, once, twice, three times, until the penguin was displaced just enough so that the skua in front could dive in and pluck the egg from the nest. Triumphant, the hunting pair would fly a dozen yards away, one carrying the egg gently in its bill before dropping it on the ground and tearing into it.

It could be hard to watch. Where there is life, death follows closely behind, like light and shadow. Energy must move through the food web: fur seals and penguins hunt krill, leopard seals devour seal puppies like popcorn, skuas pick off the penguin chicks. As a biologist, I know that things have to die and that everything can get eaten. But there's a difference between that knowledge in an abstract sense and actually watching a penguin chick being dragged away from its nest by a skua, who immediately begins to eat it, and even after a third of its body is gone, the chick is still struggling, thinking it can get away. Or watching puppies happily frolicking in a pond until a leopard seal slinks up, snatches one, and digs its incisors into the puppy's flesh. My heart still mourned for the pain, and for the powerless agony of something so young and defenseless being eaten alive.

I knew that skuas and penguins had been doing this ancient dance since long before I was born, that the skua pairs that hunted in the colonies depended on those eggs to survive and make their own eggs. That not everything on islands was some kind of beautiful symbiotic symphony—or that it was, but predation was an integral part of the music.

The emotional weight of predation implies subjectivity—whose perspective I identified with most, the egg or the hungry bird, the pup or the leopard seal. Predation caused death and also gave life. Spending time with the skuas helped me balance my penguin bias.

I watched the skuas far more than previous field techs because Matt and I were conducting a pilot study to attempt to quantify the effects of skua predation on the penguin colonies. Pilot studies test a methodology for a larger study, but don't necessarily have the rigor or data quantity that can lead to an analysis and a conclusion. As close to daily as we could for one hour, we each found high spots on which to sit where we could see most of our colonies and observe the skuas hunting. We called it the skua show. Keeping watch over the colonies with binoculars, I logged any attempted and successful predation by skuas, and which colonies they occurred in. My spot was on a high ridge, on a flat rock, at the edge of a colony.

The skua show could be slow if nobody was hunting. Sitting still in cold and wind was a different experience entirely from hiking through it, and I wore every layer I brought out from the skua shack, hands shoved deep in my pockets and clutching hand warmers. The first half hour was fine because I was warm from my hike up the ridge. Soon the wind eroded all the heat I'd accumulated, and all my muscles tightened in an effort to be smaller, to sink into the tiny glimmer of warmth in my core, to escape the wind. I shivered with each passing gust. By the time the hour was over, I was stiff and my hands were freezing, and I stumbled through the rest of my rounds and half jogged in the remaining snow back to the blissful shelter of the shack.

While it was one more thing to do, and despite the cold, Matt and I enjoyed these stints because it gave us, thinking people, time to think. The Cape prompted introspection—the long hours were physically demanding, but mentally there was space to wander. Inevitably after the skua show we submitted the topics swirling in our heads to seabird-team discussion. "How much of love is a choice?" I would demand, bursting through the door of the skua shack after one such session. "When are lies moral?" "How do gases get around the solids in our colon when we fart?"

In many ways the friendship between Matt and me is founded on the absolute freedom we feel around each other, to voice whatever we're thinking, to be whoever we are in the moment. Matt is big on communication. How we're doing, what we're thinking, how we're feeling about everything. He is exceptionally emotionally articulate, probably the result of growing up with three sisters. We could not be more different in this respect.

Because my family moved a lot, and because I went to international schools where everyone else moved a lot too, my social landscape was constantly shifting. Adjusting to these changes, plus a torturous series of unrequited loves, tuned me out of my emotions because I associated them only with a dull and persistent ache in my chest. I couldn't stop my family from moving, I couldn't stop my friends from leaving, I couldn't make my crushes love me back. The way I felt didn't change anything. I did my best to ignore or dismiss the ache. I hung around gangs of dudes equally averse to vulnerability, preferring lighthearted humor even when storms brewed inside, who took the refrain "I'm fine" at face value and let me repress in peace.

I still use humor to get out of talking about pain. Short-lived relationships in college did little to push me toward introspection, but a few patient and inquiring friends did. Matt was one of them. Sometimes I'd explain something to him, why I thought a certain way, why I felt a certain way, and I realized that I'd never actually articulated that to myself, let alone anyone else. He said once that when I opened up to him in a moment of vulnerability or, as we said "had a feeling," it was like a bird had just landed on him, and he'd have to freeze to not scare it off.

———————

Early season, skuas were establishing their territories and finding mates. Unlike the crammed, boisterous breeding habits of penguins, skuas preferred their space. Their territories could span two or three hills, and they would often opt to build their nests on the highest part of the highest hill they could find. In November, we began the joy and toil of skua rounds. During the skua show we watched the birds hunting penguin colonies, but during skua rounds we went to their nest spots and checked on their own breeding efforts.

Matt and I divided up skua territories by splitting the Cape in two: I covered the north, he covered the south, and every four days we scoured every single hill for resident pairs and potentially active nests. Since some snow was still on the ground, skua rounds required hiking up in snowshoes until I reached the rocky tops of the hills, bending down in all my layers and unstrapping said snowshoes, searching the hilltop in boots, then re-strapping the snowshoes to hike down the hill and up the next one. I saw skua pairs cawing at passing birds, defending their territory with wings outstretched. I saw skua pairs sitting together at the top of their chosen hill, feet curled up into their bodies, cozied up, feathers touching. We could see from our data set that some of these pairs had been together for a long, long time and tended to hold the same territories and nest in the same spots.

I eventually grew to love hanging out in their windy outposts far from the chaotic and pungent penguin colonies—after I got my proper legs under me. The first time I did skua rounds it took me nearly five hours to search all the hills in my half of the Cape. Finally finished hiking up what felt like every single hill in the entire planet, I stumbled into the skua shack, starving, exhausted, and excessively grumpy. Matt gave me some food and patted me on the head. I slumped into my chair and shoved granola bars into my face. He told me that it would get easier as the snow melted and the nests started appearing, and as I got stronger. It did.

During skua rounds, every four days I braved the worst of the wind from the hilltops. If it blew forty to fifty miles per hour down low, I knew it would be ten miles per hour stronger up high. On days such as these, all my skin had to be covered because of the ice hurtling through the air—any exposed patch of skin would get violently exfoliated, as if someone had pointed a sandblaster straight at my face. I leaned forward, staggered, got blown off-track by stubborn gusts that roared so loud in my ears that even when I yelled, *"HOOOOOLYYYY SHIIIIIIIIIIIIIT,"* I could barely hear myself. The adrenaline could be euphoric.

One day on my skua rounds, fighting a wall of wind up one of the tallest peaks at the Cape, I fell on my knees in a crack in the rock and let my body succumb to gravity until I was wedged between the rocks, face plastered against the soft, dark earth at the bottom. I could smell the soil and see its pebbly texture even through my face buff and ski goggles. Lying there, defeated, I escaped the worst of the wind. Only a sudden explosion of sunlight summoned me to rise again, and I heaved my weary limbs into position and trudged on.

The skuas at the top of that hill sat huddled behind a rock: a pair, feathers tousled, eyeing me warily. Almost disdainfully, it seemed. I peered into the blackness of their irises and wondered what they thought of their world, and of me in it.

I identified with the skuas' need for peace and space, far from more boisterous parts of the island. It was harder to be mad at them for stealing penguin eggs. Who was I to judge a scavenger's evolutionary niche? With millions of years of natural selection, decades of hunting skills, the ability to fly and maneuver in the windiest place I'd ever been, the cunning to steal food right from under the nose of birds much bigger and stronger than they were—skuas were kind of amazing too.

Despite my growing fondness for the sheer nerve of the skuas, it was hard to accept the ecosystem's brutality when I felt complicit. Early in the season I was checking on the nest of a young known-age penguin. I approached the colony where it nested and stood still, trying to see if the penguin would move enough to show me its egg. Young, jumpy, it eyed me suspiciously, fidgeting on its oopsie egg. I slowly backed off, but it had already been bothered and got up. This rarely happens—we stay a few yards away when looking at nests, and most of the birds pay us no mind. But the young ones could be nervous and insecure. I knew skuas were hunting nearby, but they tended to avoid people, so I lingered by the colony waiting for the bird to sit back on its egg, to make sure it didn't get swooped. The penguin, distracted, didn't yet know that the egg needed to be the center of its entire existence and wandered around peering anxiously at the world. I thought it might be better to give the penguin more space, so I walked away. I looked over my shoulder a minute later just in time to catch a skua snatching its egg.

"Fuck!" I growled at myself. *"God fucking damn it!"* Known-age nests were important sources of data, but it was more than that: I couldn't help but mourn the young penguin's precious egg and wonder if it had been my fault—if I hadn't approached to check the nest, would the bird have gotten distracted? If I hadn't been there, would the young penguin have hatched and raised its single egg successfully? Or maybe that penguin wasn't ready for parenthood. Maybe in all its fidgeting it would have lost the egg anyway. Maybe the egg would have hatched but the chick would have been snatched in one of its parent's distracted moments. I would never know, but I couldn't help but blame myself. In science, a discipline that claims to be objective, our job was to observe as carefully as possible. Any direct impact I did have on the landscape was negative—disturbance, stress, contamination.

I often thought of my days at my university's farm in Southern California, where I studied the way Tongva communities sustained themselves by enhancing the natural abundance and biodiversity of the LA Basin. I thought of the synchrony different societies had developed with ecosystems across the world, where human impact could lead to richer soils, healthier plants, a thriving landscape. The farm, which was more like a large garden, was an invitation to step in and get involved. Get my hands dirty. I loved the feeling of nurturing growing things, enriching the soil in which they grew, sheltering them from the elements, watching the leaves emerge and the birds flock to the thick greenery.

Right before Antarctica, on Midway Atoll, I worked in habitat restoration, which was a lot like gardening too. The goal of the program was to restore native vegetation to the island ravaged by military activities and a vast encroachment of invasive weeds. Deep in a humid greenhouse, I coaxed leaves from cuttings and calculated the best way to support biodiversity. It was such a clear way of improving the birds' living conditions. The crew and I planted native species that sheltered albatross chicks as they grew and stabilized the soil in which petrels dug. I lived on the gratification of playing a small part in the development of a healthy ecosystem.

In Antarctica, there was no such option. I could only minimize my presence. In Antarctica, I could not physically interact with the .environment in a way that improved it. At least, not in a way that was immediate—through CCAMLR, the science would feed into policy and lead to protections, theoretically.

I had to stretch my concept of impact to accommodate the work's outcomes. Sometimes, scaling a hill, I imagined myself as a pinprick on a map that slowly zoomed out: the rocky hilltop, Livingston Island, the South Shetlands, the peninsula, the continent of Antarctica, the entire Southern Ocean, and the earth, spinning as it hurtles through the cosmos, one speck in a vast and infinite darkness.

6.

Early December

was working in the chinstrap penguin colony on Ridge 29 when a helicopter touched down on El Condor, the hill behind camp. I heard the whirring of the propellers from the penguin colonies, and both the penguins and I looked up into the sky, trying to find the source of the noise. The sound of the helicopter was clear but distant from my ridge on the northern edge of the peninsula. It had come from the Chilean base on King George Island, fifty miles to the east to deposit two people and go promptly on its way, cutting through the air with whirling blades. Just like that, the population of our peninsula increased from five to seven.

When Matt and I got back to camp, we greeted our new coinhabitants: Renato, who was Mike's PhD student, and Federico, his technician. Renato was studying female fur seal foraging patterns under Mike's supervision and analyzing the seal data from the Cape as part of his PhD on the foraging ecology of Cape Shirreff's antarctic fur seals. Part of this work was digging through fur seal scat for krill carapaces. To help with this lovely task he'd hired Federico as his technician.

We'd known they would soon arrive, but the unpredictability of the weather window meant we hadn't known exactly when that would be. They would be staying in the Chilean huts only about fifteen yards away, officially Doctor Guillermo Mann Base. The Chilean huts were made up of a main hut with bunks and a kitchen area, a smaller "captain's hut" with one set of bunks and a small table, and a half-buried, submarine-like orange pod, the first shelter to be established at Cape Shirreff by Chilean researchers, in 1991. Like our camp, the huts were only used in the summer.

Chilean researchers were the first to start surveying the fur seal populations at Cape Shirreff. They conducted the first fur seal population census in the 1965–66 summer season, counting what they saw on the beaches from boats in the water. Due to the logistics of accessing the remote island, surveys in the following decades were intermittent. The installation of the first shelter in 1991 was followed by the two additional huts that make up the Chilean camp. While no annual research program is run from the Chilean camp, the huts often host visiting Chilean researchers, as it is a helicopter flight away from the larger permanent Chilean base on King George Island.

Renato was tall, bearded, with excellent posture and big, dark eyes. Federico, shorter in stature, was all sharp angles, jawbone to cheekbones, with a black goatee. They were both fluent in English—Renato did his grad school studies in San Diego, taking the train down to Tijuana every week to teach salsa dancing. Federico ran a consulting company that edited and helped translate scientific papers of other Chilean scientists into English.

Renato timed his arrival to help with perinatal captures, as the data was part of his PhD research. The pinniped program monitors a selection of females throughout the season, much like our selection of penguin nests. From the success rates of the subsample, Mike could project the reproductive success of the population as a whole. Unlike with penguins, when one studies a population of antarctic fur seals, it's the females that count the most. This is because so few males manage to hold territory and reproduce, and their role in creating and raising pups is limited to insemination. If a bull bleeds to death on the beach, ten more males are vying to replace him. The number of females determines how many pups will be born each year.

Perinatal captures were the foundation of the pinniped monitoring program. *Perinatal* refers to the target animals: female antarctic fur seals that have given birth to their pup in the last twenty-four hours. Sam, Whitney, and Mike would attach radio tags, time-depth recorders, and geolocators to the fur seals we'd capture. The females and their pups would be tracked for the rest of the season by Sam and Whitney.

Perinatal captures (or simply *perinatals*) were the height of the seal season, and as the second-year seal technician, this was Whitney's moment. Whitney, in her late twenties, was a veteran of the field-tech life, having done her time mostly on islands in the northwest Hawaiian chain. Whitney loved working with animals, particularly seals and seabirds—she fit into field camps easily, with consistent cheer and a fun personality. I knew her ultimate purpose wasn't tied to remote rocks in the middle of a frigid ocean—after the season she would be applying to vet school.

Unlike penguin captures, fur seal captures required lots of equipment and the whole crew. With the penguins and the skuas sitting tight on their eggs, deep in the monthlong incubation, seabird work was in a lull just in time for us to help with the string of seal captures. Matt and I did our rounds in the morning: visiting every plot and every known-age nest to check their nest contents, note which bird was incubating, and lift tails every fourth day to confirm that the eggs were still there. Afterward we met the seal crew on the beach where they'd planned to carry out the day's captures. Whitney was our fearless leader during perinatal captures, following Mike's guidance but fully responsible for all the logistics of the operation.

The goal of the captures was to pull a female and her pup out of a harem of seals that was tightly guarded by one of the hundred-plus territorial fur seal bulls defending their territories. Once we had the female, the seal crew attached devices, measured, and took samples, then released her back into the harem with her pup. We had to take the pup with us because without its mother it was vulnerable to being trampled.

Sounds simple enough, but the captures had to follow a specific strategy if we were to get in and out safely. One person noosed the pup using a long bamboo pole with a rope loop at the end, twisting it to tighten the loop around the puppy's middle. As the pup was pulled across the beach, its mother sprinted after it, away from the harem, and another technician jumped out from behind a rock and threw a net over her, pinning her flippers to her side. Meanwhile, two other people were stationed around the harem with huge bamboo poles, defending the crew from the territorial bulls, who don't want to let their females out of the harem, plus any peripheral bulls hoping to get a quickie in with an escaping female. Once the female seal is netted, two people carry her over to where we'd set up the equipment.

Mike was the puppy nooser, as this was the role that required the most attention to harem dynamics and started the whole capture off. Sam was appointed to net the female because, according to Mike's logic, Sam's history with ball sports gave him the necessary speed and coordination. Matt and I, as the penguin crew, were on bull poles, deterring the furry beasts from charging at our crewmates. Matt, as the second-year penguin tech, was assigned the bull lording over the harem, since Matt had worked a bull pole before and the territorial seal was the bigger threat. I was on the periphery, minding the bulls searching for an opportunistic chance at mating. Whitney crouched nearby with a pole as the all-around facilitator, ready to jump in to whack a bull, alert us of unusual harem activity, and help Sam carry the female over to the capture area. Renato's role was essentially to milk the fur seal females, massaging out a few drops of milk that he could then analyze when he was back in the lab. He and Federico were there to lend a helping hand with data collection, hauling gear, and the occasional additional bull-pole defense.

At the beginning of the capture, we silently sneaked in with our equipment. Mike noosed the first pup, and once it was secure and he was ready to pull, he made eye contact with us to make sure we were ready. Things happened quickly after that. I minded any other balls of muscle and testosterone that might be thinking about getting in the way, although usually the sight of me standing between them and the seal with a huge pole was enough to dissuade them. When they did rush the female or the people nearby, I jabbed at their flippers and necks with the pole as hard as I could. It made me feel like a prehistoric humanoid, spear locked in epic battle with fur and tooth.

Once the female was netted, Matt or I grabbed the pup while Sam and Whitney carried the female over to our setup, about twenty yards away. Sam, Whitney, and Mike hooked up the female to a gas anesthesia machine with a cone over her muzzle. Once she was asleep, they took measurements and glued an instrument to her back.

In the early years of the Cape Shirreff program, Mike worked with a veterinary scientist (who later went on to be the director of the Australian Antarctic Division) to develop the gas anesthesia methods that the program still uses today for antarctic fur seal females. Whitney, the future vet, operated the machine, attentive to the seal's breathing and heartbeat. In this delicate procedure, too much and you could kill a seal, too little and it would be wide-awake and not happy at all at being poked and prodded. But Whitney was more than capable: she was organized, attentive, and precise.

While the seal crew worked over the females, Matt and I carried the pup a short distance away, took samples (whisker and fur) and measurements (length, girth, weight), and, in essence, babysat it. The pups were usually less than twenty-four hours old during the captures and were squirmy balls of cuteness. They had huge black eyes and whiskers that stuck out on either side of a wet nose and were woefully uncoordinated. I'd give the pups a good sniff, to better choose a name, of course. Definitely not because their fur was soft and warm and I liked sticking my face in it. Some were feisty, usually the females, and growled and flopped their heads around, tiny milk teeth bared. Some were just trying to figure out what was going on, and they climbed all over us and poked around with their little muzzles, searching for the place where the milk's supposed to come out of. The fur seal puppy from the first capture fell asleep in my lap, eyes moving behind her eyelids during her puppy dreams. We named her Narwhal. She was adorable.

Naming study puppies is a hallowed duty of the penguin crew at Cape Shirreff. Once we decided on a name, we bleached a letter or number on the puppy's back so it could be identified when its mom was out at sea. These marks disappeared with their baby fur (lanugo) when they molted in a few weeks.

I'd heard about minding baby seals from Matt. His first season at the Cape, the year before, had been stormy and blustery. He didn't get a view of the mountains or a clear sight of the Cape for the first month, covered as it was in constant precipitation. By then, in his midthirties, he'd already been looking for a more permanent kind of life, but still he couldn't pass on applying for Antarctica and missing out on the opportunity of two seasons on the most remote continent in the world. In the middle of his first season, he told me he'd had moments when he'd be battling the wind and snow in his penguin colonies, so far from everyone, in a camp with no one he knew, and wonder, What the hell am I doing here? What is actually going on?

We all have these moments sometimes in fieldwork: a sudden wave of perspective. For me, as with Matt, it could be spurred by misery—the question becomes more like Why am I putting myself through this? Later in Matt's first season, during perinatal captures, he wrote me an email about a moment he had babysitting seal pups on the beach. This wave of perspective stemmed from wonder—there he was, on a beach in Antarctica, getting paid to hang out with a day-old antarctic seal. What even was life? How did he get there? From the many emails we exchanged over his first season, that moment stuck with me. And there I was, with Narwhal's wet nose resting on my thigh, barely out of her mother's womb, this tiny creature I was charged with minding, and the same wave of wonder swept over me. What even was life? I thought. What was I doing here? How did I manage to put myself in this situation?

Once the seal crew was done working up the female, they disconnected her from the gas anesthesia machine and put her in a large plywood capture box, which allowed the females to wake up in a dark, safe environment, away from strange bipedal humans and aggressive males. That was our cue to bring over the puppy. Mike scratched it on the head, Whitney called it a "nugget," and Sam peered into its face and asked pointed questions.

After the female had a chance to recover a bit in her dark shelter, Mike emulated the female's call to get the pup to respond. He was eerily good at this after decades of practice. Hearing her puppy call back helped the female emerge from the fog of sleep. When she replied, that signaled she was awake enough to defend her puppy from males and other females, and therefore awake enough to be released. It didn't take long. After we heard her groggy call, someone carried the puppy over to the edge of the harem, and two or three others carried the box. We planted the puppy on the beach, took off the lid of the box, and tipped it over. The female came bounding out while we retreated quickly with the box, out of sight, in the other direction. We made sure the female reunited with her puppy before picking up and moving on to the next harem. The female, a bit disoriented, made nose contact with her puppy and snapped at the bull that approached to see what was going on.

After an afternoon of nonstop captures, Sam and Whitney processed all the samples in the fur seal lab, a small enclosure on the other side of the workshop, then spent hours more inputting all the data, only to drag themselves into bed and do it all over again in the morning. The goal was to capture and process thirty females, and usually it took about two weeks, or ten days if we were doing well.

I could already tell that Whitney would be a great vet. I wondered how many other vet school applicants would have put a seal under gas anesthesia on an island in Antarctica or banded hundreds of albatross on the northwest Hawaiian islands. Whitney was an excellent field tech, but I could see the vet in her in the way she thrived during perinatals, focused and beaming. I envied her sense of direction and her certainty that the next step she would be taking was right for her. I wondered if that also came from time. I wondered if all I needed was a few more years to figure out what I wanted out of all of this. I certainly wouldn't be a vet—I am not nearly detail oriented enough to make it through medical school.

Despite the glow she seemed to take on during captures, I wondered if any of this was hard for her. If there were parts of it that just absolutely sucked. In her tenacious cheer, she was hard for me to read. Sometimes I felt as if I were talking to a personality she summoned for us, and it threw me off. I knew that she took an hour to herself in the morning, waking up early before the rest of us did. I asked her once about it and she said something about exercises and meditation and changed the subject. Throughout the day, she was always first to do the dishes after communal meals, to take out the trash, to clean up, and even when we protested, she said she didn't mind. I knew that it was true. Whatever her morning hour contained, it seemed to be the source of her not minding about the dishes or her soggy mattress or the long hours. I wanted a glimpse of the Whitney behind the smile, but she didn't reveal it easily. If she had been grumpy, sullen, or snappy, I would have felt that I knew her more. But I also knew that in a world with no privacy, some people build the walls in their heads to better deal with other intrusions.

Sam and Whitney, who spent as much time together as Matt and I did, had clearly grown close, developing their own language and mannerisms. The intensity of perinatal captures turbocharged the inside jokes. When people click and friendships solidify, the whole crew is more comfortable, more at ease. Friendship, like love, emanates.

Sam and I had our own commonalities: we'd gone to college in the same town, we were both avid readers, and we were both learning about this whole world for the first time. Sometimes after everyone else had gone to bed, Sam and I stayed up reading sci-fi books or whispering across the foldout table about said sci-fi books. Sam was into CrossFit, coding, baseball, seals, and fantasy books and openly enjoyed listening to Taylor Swift. I loved the way Sam was simply, and always, himself. I usually felt that I had to hide my tenderness, the parts of me that loved cheesy pop music and romance novels, to be this tough and carefree person, unbothered by the world. Sam's easy honesty always brought me back down to earth and invited me to be cheesy and tender along with him.

I did not feel close to Whitney in the same way, but as the only two women in camp, we had a natural alliance and our own secrets. A month into the season she let me know that she was stashing our favorite brand of energy bars in the "feminine hygiene" medical tote, tucked in among the tampons. It was the safest place in camp to hide things from men.

I admired her positivity, her work ethic, her keen eye for detail, and her sense of humor. I admired her resolve and wished I had the same crystallizing picture of my long-term goals. She'd done more field seasons than I had and somewhere along the way figured out how to connect fieldwork to the next steps of her life. If I could only be so lucky.

The iceberg split during perinatals. We were eating waffles because it was Sunday. Sam heard a distant, thunderous crack and rushed outside to find that the massive tabular iceberg to our east was now two hunks of ice, rapidly drifting away from each other. Many forces act on an iceberg once it is afloat—rain and melted snow gathering in low spots and trickling through any cracks, water lapping at the base and melting it from the sides and from below, and stressing by the natural turbulence of the Southern Ocean. It's normal for icebergs to split and flip over, but it was still unsettling if you had gotten used to their shape in the landscape.

The night the iceberg split we played hacky sack with a balloon for two hours ("Now just with your butt and elbows!" "Now just knees and face!") and probably laughed a little too manically and swung a little too hard. I lunged onto the floor in a dramatic attempt to prevent the balloon from touching down and tried to launch it upward with my face. Breathlessly cackling and lurching around the cabin, absorbed by the bouncing of an innocuous, near-weightless plastic balloon—we'd officially lost it, I thought.

Nearing the end of perinatals, we were like the frayed ends of a loose wire, crackling with erratic energy. Not having a day off in seven weeks took its toll, especially after ten days of nonstop captures. We only got weather days if the wind was consistently over sixty miles per hour, and daily averages had stubbornly stayed below forty-five since we arrived.

The skin from my fingers was always peeling off, my feet were leathery, my toes were perpetually numb, and my nose was crusty with sores from sun and wind. I slipped into the same grimy sheets every night and slept like an elephant seal on a warm rock. I had yet to do laundry because it seemed like a huge hassle: heat water on the propane stove in a giant cauldron, take it to the stay-wet room, throw it in a bucket with soap, and mash the filthy clothes against the washboard. Too much effort. When I felt gross, I gave my ears a good swab and I'd be good for another week.

Settling into the grime felt like molting—shedding the usual hygienic expectations for a greasier and more pungent state. My long underwear was my new skin. It was liberating to shrug off the pretense of normality. A collective eccentricity settled in camp.

Impossible to pretend to be a normal human being when I had to make sure to contribute to the outhouse bucket in the morning or risk having to run to the intertidal below the penguin colonies in the middle of the day, find a territorially ambiguous spot in the middle of hovering young seals, and poop on a wet rock with my bibs around my ankles, poking away approaching fur seals with a ski pole, while penguins arriving onshore stood and stared at me. Another moment of perspective: What is my life? The few times I had to resort to an intertidal poop there were always a couple penguins that seemed fascinated by the whole ordeal, coming close to peer at me and my naked ass. I imagined bafflement in their expressions and was surprised to feel defensive. Don't look at me! I thought, having observed penguins in their every intimate moment for months. Oh, how the tables can turn.

At the end of perinatal captures, Whitney and I dyed our hair with the leftover bleach we used to mark letters on fur seal puppies. As we stood outside the fur seal lab with plastic wrap over our shoulders, she lathered the lower half of my hair and I covered a chunk of hers. We washed it off with frigid water and sported our new blond streaks, just like the puppies we'd marked.

We all had our own ways to blow off steam. Whitney made up spontaneous operatic songs about the frustrations of the pinniped database. Sam did his stretches outside and walked across the deck on his hands. Matt stared at spots on the wall for hours while people bustled around him, then disappeared into his bunk at 7:00 p.m. I hung upside down from the pull-up bar and wandered down to Little Chile to hang out with Renato and Federico after dinner.

If there were waves of social energy in the evenings, the American camp would peak first, and Little Chile, about twenty yards away, would follow late into the white night. Sometimes Sam came down with me to hang out, and we switched to English but were no less animated. Federico had a keen mathematical mind, while Renato was more of an artist. When Federico decided to make a snowman to send a picture to his wife, he made a sketch like the schematic for a building and wrote a list of items he would need: carrot, shovel, olives. Renato and I found it hilarious. Reveling in the simple grandeur of the Cape, but cognizant of the drawbacks, Renato asked me one night if I could be a field tech for the rest of my working life. I told him I'd have to live it out to have the answer. The truth is that I didn't know.

I loved so many things about fieldwork. My focus wasn't pulled in a thousand different directions by colors and cars and people and a small computer buzzing in my pocket; nothing in the landscape was designed to demand attention, so I was eager to pay it. I could settle into an inner stillness and foster contemplation, cut off from the jolts of dopamine I'd got so used to with a functioning phone. There was no possibility of getting away, no driving off for the weekend, no internet to get lost in, no world apart from this one. It is ironic—islands have always been symbols of escape, but what I found most meaningful was the impossibility of escape.

I knew I wouldn't be able to do this forever—pooping in buckets, living in grimy clothes, nursing a weather-beaten face, hauling gear all day. Someday the adventure would wear off. I could see the clock all around me: Whitney off to vet school; Matt on his way out of fieldwork, still trying to figure out what that would look like and what it would mean; Renato finishing his PhD; Federico on a brief stint away from his steady and established life in Chile; Mike nearing retirement. People usually built their lives around something else after fieldwork—but not all people. I'd met field techs who had been in the game for decades, happy to live on isolated islands, more comfortable in field camp than anywhere else. I wasn't sure my mind was at peace enough for that. I loved the work, but I knew that collecting data would never feel like an end point—I was too hungry and restless, ever eager for novelty and a good challenge.

Fieldwork is like a cross between manual labor and project management, mixed in with facilities maintenance and the interpersonal diplomacy necessary to navigate various personalities crammed together in a remote camp. It was absorbing and fulfilling, but I often felt my mind wander, be it toward the research, the way ecosystems shaped our cultures, or philosophical musings. My work was bound within camp protocols, and while I made decisions around how to do it (timing, weather, gear), I didn't make decisions about the work itself. Field seasons, for all their raw poignancy, could make me feel that my brain was underutilized. I knew the program's leaders—Douglas Krause, Mike Goebel, and Jefferson Hinke—exercised their impressive intellects through their roles as lead researchers, but I wasn't sure yet if my own mind would bloom within a scientific context. I just knew that I would eventually need something more.

7.

Mid-December

On a bright, clear day we sprawled on the deck, waiting for the ship. Mike's time in camp was over. The research leads trade off field duties: one opens camp and one closes camp, with the tech crews holding it down all season. The familiar *Laurence M. Gould* would be picking Mike up on its way north from Palmer Station on one of its regular summer provisioning trips. Mike would head back to San Diego, his home and the headquarters of NOAA's antarctic program. Renato and Federico sat on the deck with us, Renato sipping maté tea from a gourd and bombilla. Soon the familiar broad orange shape lumbered around the coast toward us. The captain greeted us on the radio once the ship was a mile offshore, and we hiked down to the beach to wait. Penguins zoomed through the shallows and hopped up onshore, shooting us apprehensive looks.

When the Zodiac pulled in, it seemed like a spaceship from another world. I was deeply attuned to the four people I'd spent every day and night with for months, and encountering new faces felt like a barrage of information and uncertainty. Matt and I did our best to seem busy loading the trash and empty propane tanks on the Zodiac, avoiding eye contact. Sam and Whitney were far more sociable, engaging the strangers effortlessly in chatter. By the time Mike was a receding dot on the Zodiac, it was already 6:00 p.m.

We headed back to camp, just the four of us, feeling like kids whose parents had just left them alone for the weekend. Though nothing about Mike's personality felt repressive, his role as an established researcher and functionally our boss meant that it was still oddly liberating to have camp to ourselves.

After getting back to camp after Mike's departure, we still had daily rounds to do, delayed due to ship operations. I'd waded into the water to help push the Zodiac off a rock and was soaked to above the knee. After changing, exhausted, I procrastinated along with Sam, Whitney, and Matt, watching the fur seal puppies fighting over the pile of door covers near the deck, until it was late and we couldn't put off marching into the field any longer.

After a few fistfuls of food and a change of clothes, I was off to work at 7:00 p.m. The sun was low and a thin fog diffused the western light. Streams from melting snow gleamed gently, running down the dark earth. I had never been down at the penguin colonies this late. I picked my way across the flats and along the ridge toward the skua shack, boots sinking into the saturated soil. Water sang from all directions—the roaring of the sea, the rushing of the streams, the dripping of snow onto rock, all draining down, down, down into the anonymous oblivion of the ocean. I walked across the beach strewn with whale bones, and the fur seals looked up at me with sleepy eyes. Moss, growing in scattered patches on the ground, filtered light through thin green membranes.

I had never seen so many penguins at the beach—it was crowded with them, all fresh from the sea and coming back to their mates and their nests for the night. Their feet were bright pink on wet gray rocks, their feathers gleaming white and black, slick with waterproofing and shiny clean like a new molt. They stood on the rocks by the water in groups, preening, communing, looking around. A Weddell seal lounging nearby looked at me from upside down, belly up, eyes probing, penguins bustling around her, downy fur drying in the evening air. I could tell she was a female because of the nipples on her broad gray belly, and she cocked her head to see me better as I admired the patchy blues and grays of her pelage.

I felt as if I had intruded on something intimate, a secret moment that I was not supposed to see. I walked through my colonies in a daze, and I looked down on my data sheet as if I were seeing it through the wrong end of a telescope. How could this spreadsheet have anything to do with this fairyland I found myself in? One of the colonies had chicks a bit earlier than the others, and these tender, fuzzy creatures emitted high-pitched squeaks, faint like a whisper, but I heard them because they were new, and they led me to tiny velvet flippers and clumsy webbed feet tucked under a warm brood-patch cave. Eggshells around the nests hinted at the new, precious life hiding beneath the feathers. I'd known they were hatching soon, but it was the first time I'd actually seen them. I sat on a boulder by the beach and watched the ecosystem unfold around me, feeling moved and raw and tired. A hot mammal tear leaked down my cheek in silent homage.

In mid-December, after a long month of incubation, the chicks started hatching as expected. Peeking under penguin tails every fourth day, I saw fissures in the eggshells made by the force inside, struggling to unfold. Soon, out tumbled wet, squinty, hungry life. I loved peak hatch. It was my favorite period of the seabird breeding season. The young chicks were so precious before they grew lanky and awkward, or died.

The chinstrap chicks were wonderfully punctual and seemed to burst decisively from their shells after exactly thirty days of incubation. For all their crazy cantankerous energy, chinnies gave the distinct impression that they had their shit together. The gentoo chicks took longer to hatch: they started late, then dallied for two or three days, pipping one day, maybe opening a window in the egg another day, until finally the shell gave way and they broke free. This also seemed perfectly in character—gentoo adults often had a slightly vacant expression and tended to stop walking to simply stand near the colonies, looking around, as if they were people walking purposefully to another room and then forgetting why they were there.

The newly hatched gentoo chicks had small heads with equally miniature orange bills. Their down was pale gray save for a black cap on the top of their head, flippers the size of a paperclip. Most of their weight sat in a rounded belly spilling onto two giant orange webbed feet. Their gut was the most substantial part of them for most of their short childhood, and they always seemed to grow from the belly outward. I had to pick up the chicks of the known-age nests to weigh them, and in the palm of my hand they felt like fuzzy balloons filled with fish oil.

Often the chicks, lemon-size, slept so soundly in the nest they looked dead. They could barely hold up their heads, only making the effort to open their bills wide for a parent's regurgitated krill. An adult would open its bill and lurch in a characteristic pre-vomit heave, and the chick's whole head would disappear into the gushing maw. The tiny chicks begged for food with faint but persistent peeping.

Hatch seemed to draw everyone back to the colonies, including juvenile penguins, who had been born in recent years but were not yet of breeding age. Some were barely a year old and had just survived their first winter. I saw a trickle of juvies all season. It is not entirely known why juveniles return to the colonies before they are old enough to breed (at least three years old). Migratory penguin species such as chinstraps and Adélies have high site fidelity, meaning they are hardwired to return to their natal colonies for the breeding season. This instinct could be bringing them back to the colonies even if they are not yet breeding. It could also be practice—penguins, like antarctic fur seals, tend to be more successful at breeding after some time and experience with it. Older birds have honed their skills and are better at finding food and raising chicks. There might be some advantage to young penguins returning to the colonies and practicing their pebble-collection skills, finding a mate, and going out for foraging trips, having a little practical knowledge before they actually raise a chick, which is what some of them were up to when I saw them in the colonies.

I could identify the juveniles because the chinnies had a black smudge around their eyes, and the gentoos' white headband wasn't connected at the top yet. They were jumpy, insecure, and had a certain purposelessness that contrasted with the grim focus of parenting birds. I got excited when I saw the juveniles loitering around the colonies because they had just made it through the highest-risk year of their life: they'd jumped into the ocean for the first time ever, figured out how to be a penguin, and survived long enough to make it back to the colonies. I'd tell them they were doing an amazing job and I was so proud of the penguins they'd become.

Hatch was one of the essential dates to capture in a field season, but I knew that in the future the program might not rely on manual tail-lifting and nest checks to capture hatch dates. In addition to regular monitoring, Matt and I were conducting a validation study to test the effectiveness of cameras at capturing the critical dates of a penguin season. Jefferson, the program's primary seabird researcher, had designed the study to compare the data we gathered with our usual in-person methods with the data gathered through pictures: egg lay dates, number of eggs, hatch dates, number of chicks hatched, and the date the adults left the chicks alone in the colonies (crèched). Matt and I had set up the cameras on tripods facing the plots and programmed them to take a picture every thirty minutes between 9:00 a.m. and 3:00 p.m. The camera didn't have to capture all the essential moments because we could extrapolate from what we already knew from decades of monitoring. For example, if the hatch was caught on camera, Jefferson would know that the egg must have been laid about thirty days before. Jefferson had also noticed that right around the time the eggs were laid, both adults tended to be at the nest. Once the "clutch was complete" (both eggs were laid), just one bird would be incubating, as the penguins traded off shifts. From this estimate of lay dates, the hatch date could be calculated.

In April 2018, a year and a half after I first landed on antarctic shores, Jefferson published a paper about the new camera-based monitoring method based on data we'd collected. Checking the information from the cameras with our hands-on monitoring work, he found that 80 percent of nest chronologies could be estimated from adult attendance only, given that both adults are present during egg laying. Setting up cameras is simple and has the potential to massively increase the number of nests that can be monitored by the United States and other members of CCAMLR. It also removes the need for a daily check, as we were doing. With tightening budgets, the program has been turning to technology to help capture data and feed it into the long-term data set.

I was all for technology that might improve our ability to gather data and cause less disturbance to the animals we studied. But a small part of me felt that I was training my replacement, a cheaper, tripod-shaped competitor with no need for food, warmth, or shelter.

Penguins incubating eggs just had to feed themselves, but once the chicks hatched, there were two more mouths to feed and many more trips to sea. Adults went back and forth from nest to sea all day, weaving through the marine mammals at the beach: fur seal females tucked up with their puppies, groups of puppies together while their mothers were at sea, territorial bulls, the occasional hauled-out elephant seal. At the shore, in one moment the surf would be lapping on the penguins' dinosaur toes and in the next they'd be gone. If I was really paying attention,

I might get a glimpse of a blur of white feathers as they zoomed off into the depths, as if fired by a cannon. Anyone who says penguins can't fly has never seen those little torpedoes underwater. Both chinstraps and gentoos seek out swarms of krill, but gentoos also hunt krill-dependent pelagic fish. The food was nutrient rich, and the chicks grew up fast. In two weeks, they multiplied their weight tenfold, from two hundred grams to close to two kilos.

While I delighted in the chicks' tenderness, their vulnerability was also achingly apparent. They had to grow up fast so they wouldn't be hunted by a skua that flew low over the colonies, scanning for an opening. Skuas polished off quite a few recently hatched chicks, shaking them to kill them and then gobbling them up in a few mouthfuls. The skuas were getting ready to feed the hungry maws of their own chicks, also soon to hatch.

After peak hatch, penguin work ramped up significantly. Matt and I geared up to begin deploying radio tags. Radio tags, unlike bands, were attached to a penguin's back feathers and recorded whether the penguin was on land or in nearby waters, close enough to be picked up by the receiver antenna Matt and I had mounted on the shack. The tags would record trip duration: how long the penguins were gone from the colony, looking for food. Shorter trips meant the food was near and abundant; longer trips meant krill swarms were harder to find.

On the first afternoon of deployments, a breezy, overcast day, Matt and I stood by the colonies, nets in hand, waiting for an unsuspecting gentoo to waddle up from the beach. Once the penguin reached its nest, we could determine some essential information: the number of chicks in its nest (one or two), the location of the nest, and the respective sexes of the parents. Penguin sex is tricky to identify, but slight differences in bill size and chest shape are a lot more apparent when they stand side by side. Once the information was confirmed, we caught the "dirty" adult—the one that had been incubating when its partner walked up, and who was about to head out to sea to forage. In this way we minimized meddling with their chick-guarding shifts: the recently returned penguin stayed on the nest, as it would have otherwise, feeding the chicks from its new catch.

To catch penguins, we used nets mounted on poles—a little like large, sturdy butterfly nets. I threw the net over the penguin, then pinned its flippers as it flopped around inside. I pulled it out of the net by its thick ankles and tucked it under my arm, one hand still pinning its feet. Once it was secured, I walked the bird to the skua shack deck, where Matt would meet me. We first weighed the bird: we put the penguins in sleeping-bag sacks and hung them from a scale, as the penguins scrabbled around inside, ripping holes in the sleeping bags with their sharp nails. Most penguins weighed around 4 kilograms, or 8.8 pounds. After weighing the bird, I scooted it out of the bag and sat down on an upside-down five-gallon bucket. I tucked the penguin under my arm so that its belly lay on my leg and its head was behind my elbow, flippers pinned against its body, with one of my hands on its belly and the other holding its feet. Matt crouched in front of me, poking my puffy rain jacket out of the way, and carefully judged the positioning of a radio tag. We used zip ties and superglue to affix the black square to the penguin's back. The tag was almost hidden by the bird's feathers as it waddled back down slippery rocks, save for the little antenna that stuck straight out and made it look like a remote-control penguin.

Every thirty minutes, the receiver inside the shack searched for each tag frequency, logging the tags that were on land or in nearby waters, and noting those that were not. The receiver therefore captured how long the penguins took on their foraging trips, within a thirty-minute window. Trip duration, which ranged from three to forty-eight hours, was just one more clue in this ecosystem's puzzle. There was as yet no dramatic, discernible trend in penguin-foraging-trip durations, but the trend was clear with fur seals: it was taking mothers longer to find enough food to feed their pups.

Climate change is poised to have a big impact on the distribution of antarctic krill. Krill are a cold-loving species—they can only tolerate ocean temperatures up to 5°C—and there is evidence that with warming temperatures the range of krill is contracting southward. Following current temperature increases, some models predict that krill populations could largely shift south to the Weddell Sea by the end of the twenty-first century. The Weddell Sea is enclosed by the curve of the Antarctic Peninsula and is closer to the bulk of the continent. This shift would mean most krill would move away from the important penguin and seal breeding colonies in the Western Antarctic Peninsula. Krill populations could almost completely disappear from the region, currently known as the seeding area for krill populations in the entirety of the Southwest Atlantic Sector.

It was hard to know when, or if, the species we studied would adapt. Heavy science for someone who'd grown to know these species so well. Sometimes I had the sense that I was watching and experiencing something on the brink of dramatic change—I felt both grateful to see the penguins now and grief that the species I loved was in decline. I did not know how resilient they would be to these broad climatic shifts; I did not know what the Cape would look like in ten, twenty years. I felt that we were ever playing catch-up, barely grasping what the baseline was before it shifted. A myriad of dynamic forces would always be acting upon these animals. The Southern Ocean, like most ecosystems, is complex and deeply interconnected. We'd never be able to gain total clarity on how the system worked, but we could move closer to the kinds of insights that would help us manage the region better. For scientists drawn to polar lands, the stakes are high: you will probably be heartbroken.

Only the most dedicated make it through. One night, after more than a few pisco drinks, Renato came up with an idea for studying the adaptability of seals in recolonizing historical breeding grounds after their near extirpation by the sealing trade. He wondered if it felt brilliant just because it was late and he was tipsy, but he wrote it down all the same, and the next day he kept thinking about it. I saw him absorbed in his thesis, wondering what questions were left to be asked, already plotting new studies long before finishing his PhD. I wondered if, someday, after he'd tied his whole life to polar lands, catastrophic changes would unfold and he would return to collect data and measure the shifts. I wondered how it might feel to watch this world change so completely.

Renato's work at the Cape was not as mobile as mine, so sometimes he'd come out to do skua rounds with me, huffing up all the hills and peering at the landscape, camera in hand. From his work with algae, he knew the names of all the species of seaweed that washed up on the beaches and pointed out his favorite ones. Onshore, the aquatic strands sat in reddish-yellow piles, lifeless on the rocks. When the tide was low, we took shortcuts across the intertidal to look at the seaweed that clung to the shallows—a small window into their marine existence. For such a frigid place, the intertidal was full of life: mussels, seaweed, crusty coral, barnacles, invertebrates, all in pinks and dark greens and purples.

I have a soft spot for the intertidal—the stretch of rock between high tide and low tide—born from my summer on St. George in the Bering Sea. On the edges of windswept tundra, the dark boulders on the shore were covered with pearly algae, sea stars, and mollusks. Crustaceans skittered around salty pools, barnacles hung on to the dark undersides of rocks. The intertidal is an ecotone: *eco*, meaning "ecology," and *tone*, Greek for "tension." Ecotones are the spaces in between two defined ecosystems, such as the edge of a forest bordering grassland or the edge of land spilling into sea. Often ecotones are more diverse and productive than the ecosystems on either side.

I saw myself in this in-between, in between worlds and cultures, in between one life and an undetermined future. I liked the idea that edges were fertile ground, and I often thought of ecotones at Cape Shirreff. The South Shetland Islands were right at the northern tip of Antarctica, the closest part of the continent to more populated landmasses, the first part to be discovered, at the physiological edge of fur seals' range—the coldest place they could physically survive. The in-betweenness of this place was what made it so uniquely precious, but also what made it so vulnerable to environmental shifts.

––––––––––

For most of the season, the distinct bounds of the Cape were the edges of my life: the glacier to the south covered the narrow neck that tethered us to the rest of the island, and the ocean ringed all our other edges. While our small peninsula was only about a mile and a half from the northern edge to the south, the whole island, irregularly shaped, was about fifteen miles wide and forty miles long, with a bulbous hunk curving down to the east. When the weather was clear, I could make out the shapes of distant, snowy peaks.

We left the bounds of the Cape twice a season, when the crew hiked out to a far point of the island called Punta Oeste, beyond the reach of our little peninsular world, to survey skua nests. Punta Oeste lay across the glacier and past a rocky beach, up in the hills beyond, almost two miles as the crow flies. While the purpose was to check on two skua territories, the seal crew usually came along to spot any tagged seals that might be popping out of the sea beyond their typical survey areas.

After solstice, we looked for a weather window to do the hike. A Punta Oeste expedition demanded good weather. On a calm, partly cloudy morning, two days before Christmas, Whitney and Matt made the call. I was thrilled to venture beyond our familiar hills to new territory. Punta Oeste was a daylong expedition; armed with our binoculars and notebooks, empowered by science, Whitney, Sam, Matt, and I ventured forth just like a troop of curious fur seal puppies. We hiked south to Media Luna beach, named for its smooth, sweeping curve akin to a half-moon.

After a few appreciative moments taking in a *wallow* (the official term for a muddy heap of molting elephant seals), we headed toward the moraine. Moraines are huge piles of mud and talus that were piled up on the sides of a moving glacier as it plowed its way through the earth many thousands of years ago. Once on the other side of the slippery, muddy hill, we crossed a long, narrow trench to the beach.

We headed south along a beach that sprawled into the distance, near the glacier's edge, and hiked up the far hills. Breaking new land, we crawled into caves and scrambled up streambeds and crossed ridges. After sliding down a snowbank, I spotted a carcass on the beach, and we all gathered around it. Eventually we decided that it looked like a king penguin. Where did it come from? we wondered. Where was it going? The closest king penguin colonies were up in the Falklands, by Argentina, 750 miles away, and over on South Georgia, another subantarctic island in the South Atlantic, 1,000 miles away. Did it die here, or die in the ocean and wash up? Did it travel south to gorge on krill? The color of its neck feathers was undiminished by death, a flush of bright orange and yellow against gray rocks.

The ridges up from the beach were all wet mud and sharp stone, with swaths of white quartz embedded in the earth. Sam and Whitney lingered on the shore, looking for tagged seals. Matt and I tracked a skua pair across beds of soft green-red moss to their nest, tucked between large rocks, lichen tufts bursting from the crusty surface. One of the eggs in the nest was broken, and instead of a chick, moss was growing inside the shell. Long since dead, the egg was still faithfully tended by the skua pair. I was morbidly fascinated by the way the moss had found this source of nutrients and grown into it, turning amniotic fluids into photosynthesizing tissue. We poked at it and joked about having discovered the first-ever moss-skua hybrid, a miracle of adaptation.

Matt and I noted the nest contents and the band numbers of the two adults attending it. Checking on the nests at Punta Oeste was part of skua monitoring, except due to the distance we only visited twice a year: once when there would be eggs, and once when there would be chicks. I took GPS points at the nest to help us find it the following year, as skuas tend to nest in the same spot or one nearby. Matt and I measured and weighed the eggs, data that could serve as an indicator of how much energy the skuas were able to accumulate for their breeding effort. Skuas, much like the marine critters that came onshore at the Cape to breed, also depend on the Southern Ocean. While in the summer they scavenge on washed-up krill, carcasses, penguin eggs, and penguin chicks, in the winter, outside the breeding season, they are marine, feeding and living at sea.

Data safe in Matt's field notebook, the four of us gathered for lunch on the top of a hill. I popped open my Tupperware and carved into a cold mass of ambiguous leftovers, made from dried things and canned things and whatever vegetables had survived two months. We sat in a companionable silence, and I stared at the far-reaching enormity of the parts of the island that I never got to see. Cloud shadows moved across the glacier, with towering mountains beyond, distant and white. The bulk of Livingston Island, like much of Antarctica, is covered in an ice sheet. The island was shaped like a blotch of ink dropped on paper, lobed and irregular. Ice-free coastal areas include Cape Shirreff, the northern edge, where I sat, and the Byers Peninsula, at the western extreme, which we faced looking southwest from our perch on the rocky hill. While the Cape was a mere crumb at a little over a square mile, the Byers Peninsula made up a good hunk of the island—about twenty-three square miles of Livingston's total three hundred square miles. This larger ice-free landmass is a main breeding and haul-out area for southern elephant seals (but not antarctic fur seals). The abundance of southern elephant seals attracted flocks of sealers from Britain in the early decades of the 1800s. The Byers Peninsula boasts the highest concentration of historical sites in Antarctica. Like Cape Shirreff, it is an Antarctic Specially Protected Area, not only because of its history but because it is also a breeding ground for chinstrap and gentoo penguins, as well as other seabirds.

On the south coast, opposite of where I lived on the northern shore, tucked into a bay and on an ice-free strip of land, were the Spanish and Bulgarian camps, also only occupied in the summer. As on every shore on the island that wasn't a cliff, the camps stood above beaches made of smooth dark stones. The camps were seventeen miles away. The majority of the island was covered in glaciers and snowfields and gently rose up from the rocky shores, with the highest elevations in the interior, which was dotted with a few domes and peaks. The real altitude could be found in the eastern lobes of the island, where the Tangra Mountains towered from icy land, steep and jagged. Mount Friesland stood the tallest at fifty-six hundred feet, ringed with many other peaks above thirty-nine hundred feet. The Tangra Mountains loomed hazy and ethereal from camp in the evening light when it was clear enough to see that far.

Over the water, looking west, the Byers Peninsula's distant hills rose gently from the sea. Sam's ears perked at the sound from a blowhole. From the hill we peered down to the water, to find a few humpbacks feeding nearby. As their heads lurched out of the water, I could see the pink insides of their mouths, open wide to feed. The whales rose like ice to the surface, welcoming the ocean into their massive bodies, filtering krill, small fish, mollusks, and copepods from the water, harvesting life from the medium that nourished it.

It was a joy to explore unknown territory and let the land surprise me: molting elephant seals dyed pink by their seaweed bed, moss growing from fetal juices, the remains of a king penguin on a quiet beach. It was a joy to peer closely into little things and ask about big things, to wonder what was in store around the next bend, beyond the next hill. I felt childlike delight, as if the world were still mysterious and full of potential. The features of the landscape were much the same as what I was already used to, rocky beaches and sloping hills, but they were new, and that made all the difference.

As we sat together on the hill, I thought of how science pulls us toward answering the oldest questions: What is this world? How does it work? In science, like youth, not knowing pulls us forward.

I thought about the clever studies Doug, Jefferson, and Mike devised with cameras and old data. I thought about how their minds flourished within the structure of the scientific method, precise and analytical. The way they launched from the foundational CCAMLR ecosystem-monitoring protocols to conduct additional research into field methods and foraging behaviors. I thought about the way Mike talked about ecosystems, always calling for more studies, the way he could see the distribution of scat at the Cape or the health of the seals and infer bigger things about the broader ecosystem. My mind drifted to where he sat now in San Diego, working for NOAA's antarctic research program as pinniped lead, analyzing data, writing papers, and planning for the future. I thought about the gleam in Renato's eye when he thought of a new scientific question to ask, and I started to question the future I'd assumed for myself. It dawned on me that I'd never wanted to answer a scientific question myself. I'd never felt a burst of inspiration directed toward potential studies. I did not see myself in Renato, Mike, or Jefferson. I love science, I love studying it, I love collecting data in the field, I love learning what other people have discovered through it, and I'd assumed all those things were enough and would eventually add up to a career as an actual researcher, but perhaps I'd confused an affinity for a direction. Suddenly I wasn't so sure.

8.

Late December

In the early 1900s, Antarctica was the last great mystery. The pull of the unknown was especially strong during the Heroic Age of antarctic exploration, in which troops of men ventured forth into the white expanse with skis and dogs and sleds, before the advent of the less arduous "mechanical" age, which dawned after the 1920s. Few men (the first hundred years of antarctic history are exclusively male) managed the trip, and those who did often faced unimaginable hardships in the grips of the white continent. While I was mired in a world of stiff, sweaty socks and damp plywood floors, I often thought of a story Matt had told me of an Australian explorer and geologist who barely survived his long and painful journey through polar lands. For perspective.

Douglas Mawson was no rookie when he set out in 1911 on his Australasian Antarctic Expedition. He had proved his mettle as an expedition member by accompanying Robert Falcon Scott and Ernest Shackleton, both British, on their first attempt at the pole in 1909—at twenty-seven, he was smart, strong, tall, fit, and resourceful. Scott tried and failed to recruit him for his own poleward expedition, which he attempted in 1910. Mawson had other plans: he wanted to lead his own expedition to chart the largely unexplored and unseen coast of Antarctica south of his native Australia.

The expedition would be split between three bases to cover the most ground possible in the short summer-weather window: one on Macquarie Island, halfway between Australia and Antarctica, and two on the antarctic mainland, with the principal base under Mawson's command. Mawson intended to conduct a full geographic survey of the new land, complemented by ship-based oceanographic research. With the backing of the Australasian Association for the Advancement of Science, Mawson put together a crew, enlisting scientists from universities in Australia and New Zealand, including a meteorologist, a magnetician, a geologist, and a biologist. Aside from scientists, his team included a Swiss ski champion and mountaineer, Xavier Guillaume Mertz, and Belgrave Edward Ninnis, a lieutenant whose father had been part of a British expedition to Antarctica in 1875.

The expedition's rough beginning hinted at the hardships to come. The *Aurora*, the ship that carried the expedition south, struggled to approach land due to ice and weather. Finally, the crew fought their way to a small apron of land just as the autumn window closed, almost exactly on the opposite coast to where I was working on Cape Shirreff, in East Antarctica south of Australia. Mawson named it Cape Denison, after Sir Hugh Denison, a patron of the expedition. Cape Denison is the single windiest place at sea level on the entire planet. But Mawson and his crew didn't know that. They struggled to build their huts and hunker down for the winter through constant blizzards and violent storms.

Expeditions usually headed south in late summer to set up a main camp, then spent the winter preparing for the expedition, which would launch as soon as the weather warmed and was hospitable enough to allow the crew to cover ground. Mawson's plan was to overwinter, explore the unknown coast in the subsequent summer, and then be retrieved by the *Aurora* in the fall, just before sea ice formed from the coast and made it difficult for ships to pass.

Winter was long, harsh, and unbelievably windy. Mawson's men set up weather stations and logged the daily fluctuations of the extreme weather. The measure of wind is a simple thing: a turning pinwheel attached to a device that measures the speed at which the pinwheel turns, otherwise known as an anemometer. Ours today are upgraded versions of theirs, but not by much. Back then, analog dials converted the turning to wind speed, while a small digital screen does the same for us. In May 1912, in the antarctic fall, the expedition's meteorologist calculated that the average wind speed was sixty-seven miles per hour, far above a gale and just shy of a hurricane.

Eventually, the land showed the first signs of spring for the weather-weary crew. In October, Mawson's crew saw the first wildlife returning when a penguin hopped onshore. Ecstatic, a crew member gave the startled bird a bear hug and brought in into the hut, plopping it on the mess table for a toast. I must assume the poor penguin was eventually released as the nineteen men set about getting organized for their expeditions.

The summer's expeditions would be split into four parties: one going south toward the southern magnetic pole, one to explore the plateau that lay to the west of camp, one to venture east along the coast charting and surveying the shore east of Cape Denison, and a far eastern party, which would cross the longest distance and attempt to reach Oates Land, about 350 miles away. Some men would form support parties, carrying supplies and largely assisting the main expeditions. Everyone absolutely needed to be back at Cape Denison on January 15, 1913, when the *Aurora* was scheduled to pick up the crew.

On November 10, 1912, the far eastern party departed the huts, consisting of Mawson; Belgrave Ninnis, the dog handler; and Xavier Mertz, the ski champion. They pushed east through punishing winds, with three sledges and two dog teams. Despite the weather, they made good progress, covering about three hundred miles in thirty-four days. The journey included crossing the largest glacier yet known in the world, minding dangerous crevasses covered with treacherous snow bridges. One day Ninnis suddenly plunged through a crevasse and disappeared, taking with him the rear sledge and the six best dogs. As the others stood at the edge of the crack, coming to grips with their new situation, an antarctic petrel appeared and hovered over the crevasse before disappearing again into the white. The sight of another living thing in this desolate landscape, especially something so delicate and ethereal, was some comfort—Mertz believed it was Ninnis's soul bidding his final farewells.

The only thing to do now was to turn back and pray they could cover the three hundred or so miles back to the huts before supplies ran out. They had about twelve days of rations and two smaller dog teams. The journey back was torturous and delayed by the ever-present blizzards. One by one, the dogs collapsed and were butchered to feed the rest of the huskies and the two men. Mertz and Mawson saved and ate the husky livers, which are now known to concentrate vitamin A. Once vitamin toxicity was better studied, historians estimated that they took in, between them, sixty times the toxic dose of vitamin A. Sharp stomach pains, nausea, and problems with balance soon accompanied their trudging. Then their skin started sloughing off, exposing the bare flesh underneath to the harsh environment through increasingly tattered clothing. They killed the last dogs ten days into their return. After marching almost halfway back to the hut, Mertz was unable to continue, lying in his sleeping bag in fits of madness and deepening depression. Mawson, acutely aware of their diminishing stores, was obsessed with covering distance and in vain tried to rouse Mertz and urge him to trudge on. Mertz died during the night.

Mawson now found himself completely alone in a forbidding place, still one hundred miles from the huts and any hope of survival. Vitamin A toxicity ravaged his body, and skin shed from his legs, groin, and ears in sheets. Mawson struggled south, dropping anything extraneous from his load and pulling the rest behind him with a dog harness strapped to his middle. A few days after Mertz's death, Mawson sensed a new discomfort in his feet, and after camping, he removed his shoes to find that the skin of the soles of his feet had come off. Before his daily trudge, Mawson would strap on the soles of his feet. Matt emphasized this part when he told me the story during one skua shack afternoon. It was one of the most vivid details he remembered, stressing the insanity of this part with wide eyes and dramatic hand gestures: "Every morning he strapped on his feet. He literally had to tie on the soles. Of his own. Feet. Every morning." To give his feet a break, Mawson sometimes got on his hands and knees and crawled for miles.

Halfway across a glacier, today named Mertz Glacier, he suddenly plummeted through a snow bridge into a crevasse. The sledge held firm, and he dangled on his rope, four yards down, between two smooth walls of ice. Mawson thought it was the end and doubted he had the strength to haul himself back up to the surface. He had his penknife in his back pocket and considered cutting the rope and falling to a sudden death. But the thought of the meager food he still had stashed on his sledge rallied his spirits—he could not let that food go to waste. He hauled himself hand over hand up the rope, while the skin on his palms abraded and fell off. Finally reaching the snow bridge, he leaned on it, only for it to give way, and he plummeted back down into the crevasse. I cannot imagine this moment. He had no skin left on his hands and was already at death's door. Somehow, he managed to pull himself up and passed out beside his sledge. He camped there that night after coming back to his senses, resting, writing, and recovering. Eventually, he struggled on, finally crossed the glacier, and ascended up the slope to the polar plateau. Weeks after the soles of his feet had come off, he discarded the skin because it had hardened and was hurting the exposed flesh beneath.

All the other parties had returned on time to the huts at Cape Denison and were growing increasingly concerned about the disappearance of Mawson, Ninnis, and Mertz. The ship lingered offshore, waiting, the agreed departure date almost three weeks past. When the *Aurora* was forced to depart due to a closing weather window, a group of five men were left behind as a search party for the missing expedition. Starved and half-dead, Mawson finally staggered within a mile of the huts and saw three men bent over something by the shore. He waved to them and, once he was seen, collapsed by his sledge. When they reached him, they did not recognize him: he had deep open wounds on his face, his frame hairless and skeletal. Little was left of the man they had seen depart from camp months before. They brought him back to a hut and nursed his wounds and his hunger. He had missed the ship's departure by just six hours. Stuck at Cape Denison for another winter, he rested and recovered from his ordeal.

During the second winter, the Cape Denison party established radio contact with the world through a relay station on Macquarie Island, halfway between Australia and Antarctica. This was the first radio contact established from Antarctica to other continents. The previous winter communications had been attempted but were unsuccessful due to the wind and other technical issues that the crew was now able to resolve— until the radio mast was again brought down by wind in midwinter.

The expedition as a whole was huge in scope: the four sledding parties that departed from Cape Denison, the additional base established six hundred miles west of Cape Denison, the party on Macquarie Island, and the scientists aboard the *Aurora*—all gathered extensive information on this new region, describing in detail the continent's extreme conditions and unique species. The expedition's detailed magnetic, meteorologic, geologic, geographic, and biologic observations were published in long reports over the next thirty years. These efforts contributed to Australia's eventual territorial claims over that particular "slice" of the continent.

Later in life Mawson led other expeditions to map the antarctic coast south of Australia and New Zealand, was knighted, and became a professor. The glaciers in Commonwealth Bay are still called the Ninnis and Mertz Glaciers. Australia's antarctic base closest to Cape Denison was named Mawson Station.

_____

While men dominated the continent's early history, the continent itself was often feminized in the records kept during the Heroic Age. In *Antarctica in Fiction*, Elizabeth Leane likens the continent to a devouring womb, with deep fissures that consume men, as Ninnis had been consumed, as well as a feminized body: with "its ability to swell and shrink and to break off—or more specifically 'calve off'—parts of itself, the Antarctic shares the abject qualities of the maternal body." A poem by Ernest Shackleton, written when he accompanied Scott on his first attempt at the pole in 1901, describes the Ross Ice Shelf as "mother of mighty icebergs . . . the great grim giants you wean / Away from your broad white bosom."

In the nineteenth and twentieth centuries, novelists also explored Antarctica as a vast, unknowable otherness—as a depository for the fear and darkness we harbor in our subconscious. Horror writers used Antarctica as a Gothic landscape, fueled by the very real horror and hardship experienced by explorers such as Mawson. Novels such as Edgar Allan Poe's *Narrative of Arthur Gordon Pym of Nantucket* (1838), H. P. Lovecraft's *At the Mountains of Madness* (1936), and Murray Leinster's *The Monster from Earth's End* (1959) were set in Antarctica but with imagined twists: the pole as a portal into the interior of the earth, alien monsters buried in ice, supernatural creatures, giant malformed animals and zombies. The Gothic is written into the continent's very geography—islands near Livingston bear names such as Desolation and Deception Island. Sitting on the bottom of the globe, Antarctica is described by Leane as a literal underworld.

_____

Unpopulated until recently, Antarctica is a canvas on which our collective psyche can be painted. For Matt, Sam, Whitney, and me this included celebrating holidays despite the landscape's indifference. One of my favorite parts of a remote field season was that it was completely up to us what and how to celebrate. Inevitably traditions had to mold to fit our particular context.

Before I knew it, New Year's was upon us. Since there were no local customs, I coerced everyone into following mine. My family had accumulated New Year's traditions from every country we'd lived in, so on that holiday we had to appease a laundry list of superstitions if we were to have a good year. In new countries we added to these traditions, but in Antarctica they had to be amended. Sam, Matt, Whitney, and I threw coins out the window for good luck (a Spanish tradition), then I went out and picked up all the coins from the snowbank because littering is against the Antarctic Treaty. I cleaned the camp of evil spirits by waving a pot of water over everything—water that we then used for dishes. I rehydrated a bunch of raisins for the custom of starting the New Year choking on fruit (also Spanish): you must consume one grape per second for the last twelve seconds of the year—the twelve grapes of luck. All good Spaniards follow the same ritual, ideally to the timing of bells that ring in Madrid's central square. In the absence of the bell, I banged a large metal bowl with a soup ladle to signal the seconds according to Whitney's watch, and we all stood on the deck shoving rehydrated raisins into our mouths. We didn't have gold for our champagne glasses, so I cut some golden foil from a camp decoration instead. It was another good-luck ritual passed down from my parents, the national origin of which I could never keep track of. If I get struck by lightning someday because of my blasphemous ritual modifications, let it be known that I did my best with what I had.

At midnight we toasted with champagne and finished chewing our wet raisins. We immediately took off to run a circle around both camps with drybags, a version of running around the block with a suitcase if you want to travel, which I have also done on every block in every city in every country I've ever lived in. My mother, ever superstitious, collected these traditions like lucky charms on a bracelet.

Matt opted out of our loud celebrations and sat on another part of the deck, watching the waning light with a contemplative look on his bearded face.

Beyond penguins and ice, beyond a dark underworld, there was still space to indulge in everything that made us human: the tug of our cultures re-created in inhospitable lands, the comfort and messiness of relationships.

One night, after everyone else had gone to bed, I asked Renato to teach me salsa. He took my hands and swayed me around the room, instructing me on beats and steps. Halfway through the lesson, the speakers ran out of battery power and the music died out. We kept dancing, sashaying around the kitchen, Renato softly counting out the beats with the metronome regularity of a seasoned teacher, one, two, three, and a five, six, seven, and twirling me, damp socks hanging by our heads, boot dryers in the corner, a faint smell of mold permeating the cabin air. It was late, and a light dusk was falling outside the window as we stepped across the old boards. I liked our long talks and our shared impulse to philosophize, and I'm a sucker for men who dance. Our coexistence on the island was already so tightly woven that to slip into other kinds of intimacy seemed easy, and natural—we kissed on a chair in the hut while the wind and the seals howled outside.

Romance in the field can be a huge risk, but also a welcome joy in a lifestyle that leaves little room for relationships. Hard to date when you're always running off to the next island. Part of Cape Shirreff lore is the story of the union of two field techs who worked at the Cape three, four, five seasons in a row, fell in love, and are still together today. It's difficult to maintain a relationship otherwise, with one person always off working somewhere far away, but it can be done. Whitney was the only one among us with a partner back home, and she often disappeared to the fur seal lab to use the satellite phone.

Everything about our lives was tied to the job. The perspectives we brought with us changed to accommodate the stark necessities of a

Spartan field camp in Antarctica, and of the land itself. It was not unlike the international schools of my childhood, where the place in which I lived was nobody's home, exactly, but nevertheless created communities with a strong sense of place and context. My own cultural perspective, by necessity, has always been "when in Rome." But what if there is no Rome? To build something deeply human in a world that seemed so inhuman was a special thing. It's in our nature to reach for the unknown, but we can't survive on uncertainty alone; something always pulls us back to ourself and to one another. Something like the ritual of a holiday, the comfort of a friend, the closeness of a lover, the promise of a meal. Mawson decided to live because there was still food in his sledge. That moment was not about grandeur or wonder or the vast unknown. The human body is remarkably resilient, but it has its own demands. You cannot live off wonder alone.

We need a base from which to launch. Like a lapping wave, we reach, and retreat, and reach, and retreat. It takes all the mundane, human rituals of a daily existence to enrich the soil from which moments of poetic insight can bloom. Mawson's labor was long, painstaking, tireless, and necessary. But thanks to that labor, he had experiences like this:

The tranquility of the water heightened the superb effects of this glacial world. Majestic tabular bergs whose crevices exhaled a vaporous azure; lofty spires, radiant turrets and splendid castles; honeycombed masses illumined by pale green light within whose fairy labyrinths the water washed and gurgled. Seals and penguins on magic gondolas were the silent denizens of this dreamy Venice. In the soft glamour of the midsummer midnight sun, we were possessed by a rapturous wonder—the rare thrill of unreality.

CRÈCHE

9.

Early January

As the chicks grew, we passed the season's halfway marker. Early January saw Matt and I deploying radio tags on chinstrap penguins whose chicks were a week old, tiny in the nest but quickly growing. I was still checking for new hatched chicks on daily rounds and noting any that might have been missing, likely carted off by a skua. The chicks would be sheltered by their parents as they grew, until they were old enough and large enough to stay warm by themselves and be safe from skuas. After about four weeks after the hatch, penguin parents would no longer be on the nest every second of every day, instead leaving the chicks alone in the colonies. This was called crèche.

After crèche, the chicks would start growing their real feathers, better for staying warm in the ocean, rather than the fluffy down that kept them warm on land. Chinstraps and gentoos had different life histories after their chicks feathered out: chinstrap adults would abruptly stop feeding their chicks, who, fully feathered, would head to the water's edge, leap into the surf, and begin to hunt for themselves, migrating along the continental shelf. Chinstraps spent all winter at sea. Gentoos, on the other hand, provisioned their chicks over a longer period. Feathered gentoo chicks grew independent gradually, going on short trips to sea while still receiving supplemental meals from their parents, until eventually the chicks transitioned to independence and fledged. They'd spend the winter around their natal colonies.

But crèche was still a while away. The chicks were small but growing more coordinated: they could lift their heads, and their calls became less squeaky. They started moving around in their nest, testing out the anatomy of their growing bodies and developing the mannerisms of penguinhood.

———————

In early January, camp received a reprovisioning run along with the arrival of two additional members: Douglas Krause, a seal researcher and our new crew lead, and Jesse, a NOAA Corps logistics member who also helped with the seal research. I'd met both Doug and Jesse when I stopped by San Diego before the season started.

Contact with outside people brings viruses and germs. In the days before resupply, we had been drinking gallons of Emergen-C in an attempt to fortify our immune systems. Sam poured the powder straight into his mouth and washed it down with orange juice, sometimes water, once milk ("Never again," he'd sputtered). On the day of resupply, we reminded each other, *"Don't touch anybody."*

Researchers and support staff on their way to Palmer Station came onshore in the Zodiac to help us load our trash and unload produce and other gear. They had likely flown from the United States to Chile and spent a few nights in Punta Arenas, as we had, waiting for the ship to depart. Palmer Station, like any other base in Antarctica, was chock-full of scientists studying ecology, glaciology, geology, microbiology, and a number of other fields. Research peaked in the summer during the breeding seasons of marine species. On the way to their duty station, Palmer Station personnel could volunteer to help move gear on and off the beach. In return, they got to land on a protected island and visit camp. Three or four visitors walked through camp, peering discreetly at our plywood walls covered in maps and cards and drawings, at our rudimentary kitchen, at the table where we had dinner every night. While they never leered or were overly objectifying with their observations, I still felt a little like a specimen on display, the innards of my intimate domestic life exposed.

I tried to see main camp—the hut in which we cooked, ate, worked, and slept, the supply hut, the outhouse, and the adjoined workshop and fur seal lab—as they might: shelves mounted anywhere they could physically fit, crammed with food and totes and gear. Any open wall space was covered in vestiges of those who had come before us: cutouts of cartoons about penguins, a note in which the writer apologized profusely for finishing ("totally NOSHING") a tub of Nutella, a written menu from a celebration, small trinkets hanging from thumbtacks. An invitation card to a cocktail party at the skua shack in 2012 with an illustration of a penguin holding a martini glass. Seal drawings abutting maps of the Cape. Three solar-powered plastic flowers sat on the windowsill, swaying and bobbing when the sun was out. A wooden mobile of a dinosaur hung over the trash. Field camps are always full of ghosts. Main hut was no exception.

When the visitors left, I was relieved to slip back into normalcy, just us and our decaying totems, sagging shelves, walls heavy with history, and the warped plywood floors that I was always tripping over.

Doug was a scientist from the program who lived and worked in San Diego. He had done his PhD on leopard seals and was working with all the seal data at the Cape. With Mike's retirement on the horizon, Doug was preparing to take over the duties of lead seal researcher, alongside Jefferson, the lead seabird researcher. Jesse was part of the NOAA Corps, a uniformed service that provides operational support for NOAA's research and mission, deploying pilots, engineers, scientists, and project managers across the agency. Corps officers went through basic officer training for nineteen weeks and were then assigned to a program for three years. Jesse's job was logistics, administration, and scientific support. She would also help with the daily seal survey, in which seal technicians checked the study harems for puppies and females, recording their location and noting any other tagged seals in the area. Jesse was also tasked with generally organizing and improving camp. Back in San Diego before the season began, she had bought, packed, and shipped all our supplies. Like Jesse, Doug had served as a NOAA Corps officer at Cape Shirreff, his start with the program, before he'd gone on to work as a field technician and complete his PhD on leopard seals. They made quite the contrast: Doug a lanky tower of a man and Jesse shorter and compact. She was both stern and goofy, and always brutally honest. Doug had the easy charm of an extrovert and the storytelling skill of a drama major. They were both smart and witty as hell and gave us laughs in exchange for the physical space they were now occupying.

In addition to people, the ship brought us blessed, blessed fruits and vegetables. We went from one solitary moldy cabbage to a fully stocked vegetable haven, and every time I went to the freshies room, I was overwhelmed by the colors, smells, and vitamins. I luxuriated in the avocados, lettuce, and peppers and basked in the glow of nutritious colors and crisp, fresh food that goes crunch.

The night of resupply we had tacos for dinner. While Jesse, still encased in three layers of warmth despite being inside, pressed masa into tortillas, she asked us how the season was going.

Despite the strain of the wind, the long hours, the hiking, the hauling, and the isolation, I was having *fun*. The penguins had so much more personality than I'd anticipated. All the chicks were hatching, enough of the snow was gone that we could walk without strapping on snowshoes, and Matt and I were busy deploying radio tags and capturing hatch dates. Every day, I checked on my fifteen plots and thirty-eight known-age nests, taking the same route through the colonies every time, stepping on familiar shit-stained rocks. As the chicks grew in the nests, we would deploy recording devices that measured the diving depths and locations of the adult foraging trips. We would be collecting diet samples, weighing the chicks at twenty-one days, conducting the chick census and chick banding sweep. The bulk of our work was still ahead of us.

I loved how simple my life was, how everything I needed was contained on this peninsula, in this little hut. There would always be food, community, and purpose to my days. No traffic, no grocery shopping, no ads, no strangers, no concrete, no social media. My life was still riddled with inconveniences, but they were different from those encountered in urban places: moist boots, dirty socks, hauling heavy buckets, the cold, the inaccessibility of showers. It all seemed like a small price to pay for peace of mind. Somehow it was easy for me to just *be*. Even so, an uncertainty simmered in me—no longer attached to a researcher trajectory, part of me felt adrift, indulging in the moment because the future was harder to see.

I didn't quite know how to answer Jesse's question, how to package the experience into something that could be communicated, but "good" was the overall consensus among us.

Doug and Jesse shared the small bedroom behind the supply hut (the "old fart's room"). Our kingdom shrank with two more bodies in the mix. It was almost a curse to have enjoyed camp all to ourselves because then came the task of recalibrating. Main hut felt as if it were at full capacity. Doug and Jesse looked and smelled clean, and when we told Jesse we averaged a shower every two weeks, she arched a single eyebrow.

Our germophobia had been justified, as Jesse came off the ship sick. She sniffled and stayed inside, miserable. Doug was the next to fall. We flocked to the hand sanitizer, washing things as well as we could and praying that our health would hold. Getting sick here was a minor catastrophe, not only because of how far we were from care, but because we worked long hours, mostly outside in harsh, wet, windy, cold weather.

The work Matt and I were doing in the colonies reached a crescendo during chick-rearing. The gentoo chicks were about two weeks old and the size of a coconut, while the chinstraps were one week old, still small in the nests, like soft, fuzzy oranges. I still did my regular nest checks in my fifteen plots, as well as of the thirty-eight known-age birds, and noted the hatch dates.

After the radio tags, we also deployed rounds of time-depth recorders and geotrackers on adult penguins, all to contribute information on where and how the penguins were foraging. Radio tags, which we had already deployed, measured presence/absence data over time and indicated trip duration. They stayed on for the whole chick-rearing period, until the birds molted and the tag fell off. Time-depth recorders (TDRs) measured the diving profile of foraging trips by tracking depth (water pressure) over time, and PTTs (platform transmitter terminals) were geolocators that pinged a geographic location to a satellite when the tracker broke the surface of the water on the back of a foraging penguin. The PTTs and TDRs we'd gotten at resupply were deployed for a week only, and had to go out as soon as possible, before the chicks crèched. Sometimes fieldwork was a sea of tiny machines.

The skua chicks were also hatching, meaning that we had to conduct daily territory checks to nail down hatch dates. The skuas became extremely irritable when their chicks hatched, prone to screaming and dive-bombing. Matt and I were wrestling with penguins and fending off skuas all day and working ourselves to the bone. I had bruises all over from penguin beaks and slaps. Matt hadn't been sleeping because he was stressed-out and other people were staying up late in main hut, and I hadn't been sleeping because I'd stay up hanging out in Little Chile. We were both running on fumes. It shouldn't have been a surprise that when Matt got sick, he went down hard.

One morning after a long day, Matt didn't get out of bed. I waited for him to emerge, thinking he'd just overslept. When I checked on him, he was barely responsive. He had a high fever and a crippling headache. He couldn't get out of his bunk that day or the next and was a quiet, pained presence on the other side of his bunk curtain. I fed him soup, refilled his water bottle, and passed him Tylenol. "Water is gross," he moaned, eyes squinting, voice hoarse. "I don't want it." I sighed and threw a peach tea bag and some honey into the bottle, shook it up, and stuck it behind the curtain. Then I was out the door to write things down about birds. I rarely saw him during those days, except for his daily creep to the outhouse. The rest of us made a "Matt log" on the whiteboard, marking the times he took meds and whether he had a fever or was responsive. We tracked his health as if it were a penguin nest, diligent with our data collection and with much more at stake.

In medical situations we could consult a doctor on the satellite phone and had a large medical kit outfitted with various prescription drugs we could administer with the doctor's guidance. The doctor was part of a remote-medicine service that catered largely to cargo ships and other marine operations, where people worked in hazardous conditions far from definitive care. The doctors in the service could be working from anywhere in the United States. I called in and described Matt's symptoms. Because of his massive headache, the doctor thought it might be meningitis, which could be life-threatening.

We had few hospital-evacuation options. A helicopter from neighboring King George Island could come retrieve someone and then put the patient on a flight to Punta Arenas (King George was outfitted with a small runway). This would take two days at least, but helicopter flights were hugely weather dependent. We could get picked up via ship (ten days, depending on ship availability and the weather). Neither option was assured. In a medical emergency, we were largely on our own. As the person in camp with the most recent medical certificate, I was the primary medical officer. I'd taken the wilderness-medical-responder course before the season, and it prepared me to deal with medical emergencies in remote settings, delaying critical situations for hours or days until a hospital could be reached. If something did happen, it was up to me to make a hard call. Every day that a health crisis did not materialize, I breathed a sigh of relief.

With Matt sick, my latent anxiety over the responsibility of dealing with a potential medical emergency crystallized around the person who was most important to me in camp. The feeling of a tight twist in my stomach and a racing heart in my chest was deeply familiar to me—just like what I used to feel in my childhood when my mom took longer than usual at the grocery store, when my sibling was late in getting home, when either parent was on a reporting trip to a country in crisis. I worried and worried. I was a nervous kid, abnormally attached to my nuclear family, the only source of constancy in my life. I didn't feel capable of spending a single night away from home. I went on an overnight school hockey trip when I was eight or nine, and I remember asking my roommate for her phone so I could call my mom, and crying into the night. I couldn't handle it.

Strange to imagine that little girl, so anxious about leaving her family, when as a young adult I'd lived out of my duffel bag for years, hopping on planes to far-flung places without a second thought. I grew out of it—save for that anxiety that coils and churns in my stomach whenever I feel that someone I love is in danger or when I feel vulnerable in a big and complicated world. My thoughts circled like vultures, just as they did when I was little: What if Matt's in danger and I can't help him? What if he needs to be evacuated but the weather is bad, the helicopter can't come, and there's nothing I can do? What if I have to make a critical decision and make the wrong one? What if something really, really terrible is wrong with him and we're stuck all the way out here?

I liked the isolation until access to modernity became a matter of life or death, even if it was just in my head. Suddenly the island wasn't so fun.

I switched into overdrive and became a one-woman seabird team. I checked on my nests, checked on Matt's nests that were about to hatch, went on my skua rounds, and checked Matt's skua nests that were about to hatch. Everything seemed unbalanced. I felt as if I were missing a limb. On the second day of Matt's illness, I'd just finished checking on one of his territories and was sitting on the side of a hill, facing the sun and trying to photosynthesize, thinking, It will be okay, it will be okay. I was slated to meet the rest of the crew on the beach nearby for a round of fur seal pup weighs.

I caught a flash of color in my peripheral vision and turned. Someone was coming over a ridge, from the penguin colonies to the north. I peered through my binoculars, perplexed. As the person approached, two more people rounded the hill I was basking on. Confronted with two people in bright clothing whom I didn't recognize, I was convinced I was seeing things, but then one of them waved and held out his hand. Just by his *"Hola!"* I knew without a doubt that they were Spaniards. I had heard of a small Spanish base on the south shore of Livingston, across the glacier, which seemed so far away it might as well have been on another island, or another planet. The base was opened in 1988 and named for the king of Spain at the time, Juan Carlos I.

I felt like an eighteenth-century explorer, encountering other people on their stretch of barren land, before phones, before the internet. I felt as if I should be wearing a knife at my waist and a leather cape that billowed in the wind.

The Spaniards asked me about the Chilean program because my Spanish sounded South American and they assumed I was Chilean. I told them no, I worked for the United States, but I was actually Spanish, just like them. They were perplexed—I hadn't had to explain my background like this in a while, and I was still stunned by their sudden appearance. The familiarity of their lisp and harsh *j*'s threw me back to the smell of my grandmother's apartment in Madrid, her *tortilla de patatas*, hot summers at the pool with my cousins, mouth full of *jamón serrano* on Christmas, cool and bright Spanish nights. To that little girl who hopped borders like lines in the sand, too culturally spread out to relate to her own grandmother. That little girl who fretted and worried, clutching on to her closest people like a buoy in a vast and turbulent ocean.

The Spaniards were mountain guides, scoping a route to bring Spanish glaciologists to this side of the glacier. Spanish scientists were looking at the ice dynamics of the island's glaciers and sought to expand their study area by surveying the glaciers far north of their base. They'd snowmobiled their way across the glacier for hours to get to our side of the island. Renato told them they were free to take a look around camp and we bid our goodbyes, and after they walked off, I started laughing because it was so strange to have run into them, because the neat dividers that kept my worlds apart were collapsing, because I needed to do something to release all this tension I was holding in my body. I kept laughing while I chased fur seal puppies with nets across the beaches, because I was scared.

———————

To my immense relief, after three days of hibernation the man behind the curtain gradually came to life again. Usually, Matt is a solid, grounded presence, composed, thoughtful, inward. In our friendship, he is the plant and I am the bee that buzzes on waves of happy creative chaos, flying in erratic loops around him. In my attempts to hold down the fort and take care of him, this dynamic was exacerbated to an almost comical extreme. Matt was barely moving and I hadn't stopped. The next day he could hike (albeit slowly), so we jumped immediately to conducting diet samples. We'd delayed the diet-sample collection due to his absence and could not wait any longer. We had only conducted one round of diet samples before he got sick, and I had been preparing to do it with the crew, leading it on my own. I bustled around him, prepping everything, bringing the penguins, restraining them, pumping them, and then carrying them back to the colonies.

January was our busiest month. When the chicks hatched, the adults came and went from the colonies to forage so often that we had only a brief window to gather information on these trips, knowing they'd soon return to the nest. In January we did two rounds of geotracker deployments and time-depth recorder deployments, twenty-one-day chick weighs, diet-sample collection, and daily skua-nest checks to catch hatch dates (as opposed to every fourth day, as during the rest of the season). No time could be wasted.

Diet sampling was the most intense and arguably excruciating part of the season for seabird technicians. It took a lot of time, focus, and preparation and was highly technical and emotionally draining. Sampling penguin diets was the only way of literally seeing what the birds were hunting in the ocean near their breeding colonies, of measuring the size of the krill they were catching and accounting directly for their diet.

It was also the most invasive process we had to do with penguins, and we required the help of Sam and Whitney, who joined us in the colonies to help catch penguins. Penguins returning from the ocean carried krill in their stomachs to regurgitate to their chicks. We'd wait for a returning penguin to greet its mate, write down the number of chicks in its nest and the sex of the bird, then fling a net over its body, gather it up, and take it to the deck of the skua shack to be "pumped," i.e., to collect a diet sample from it.

A bladder with warm water, a plastic hose, a bucket, and sieves were already prepared on the deck. I held the penguin between my legs while Matt opened its mouth and slid the hose down its esophagus. Then he raised the bladder and the water slid down the hose into the bird's stomach. He held the bladder up until we heard a little gurgle. I grabbed hold of the penguin's feet and stood up, holding it upside down and supporting its back against my leg while Matt opened its bill and massaged its neck. Water came pouring out, along with the meal the bird had collected for its chick—usually a gush of pink krill. Matt taught me how to massage a penguin's neck to extract the food, which is really, really tricky. Penguin necks have a lot of skin and a thick layer of feathers. You have to apply the right pressure at the right angle and move your hand along just the right spot—or risk killing the bird.

Once the flow ebbed, we pumped the bird one more time, emptied it, then set it down at the edge of its colony for it to return to its nest. The birds were always dazed. In Matt's words, they looked as if "they'd just been hit by a truck," shaking themselves out to clear their head and throat.

Penguins live a harsh existence—getting hurled against rocks as they come in from the water, navigating frigid currents in the open ocean, fleeing from leopard seals, having frequent violent fights. They are survivors with a mettle that never ceased to impress me, and I knew that they could handle it. But I also knew that I was traumatizing the birds I claimed to love, even if most of the time they would be just fine. Diet sampling was the only way to get direct access to what penguins eat in order to better protect their populations. We were stressing a few birds for the sake of the entire species—but that didn't make the process any easier.

Diet sampling required calm, assuredness, and precision. We processed four birds per diet-sampling "round." Diet sampling started about a week after peak hatch for both gentoos and chinstraps, with a round of four birds every five to seven days. We were to complete five rounds for both chinstraps and penguins (forty birds total) during chick-rearing. It usually took about four weeks from peak hatch to peak crèche.

After we collected all four samples in a round, Matt, Sam, Whitney, and I sat down in the skua shack and processed one diet sample each. We kept a bottle of whiskey in the skua shack for diet processing because by the time we'd collected all four samples, we all needed a drink. I can't remember the taste of that whiskey without also remembering the pungent stink of regurgitated fish emanating from four sieves of pink slop, slowly warming as the propane heater kicked in. Processing the samples added one, maybe two hours to our days, more for Matt and me, who stayed to clean up.

To process the samples, each person picked the fifty most intact krill from the wet, pink mess that was their allocated diet sample. The fifty krill were then sexed (males have a small red dot on their abdomen) and measured for length, from black eyeballs to the tip of their tail. The measurements we took were a direct way of getting information on the swarms near the Cape's breeding colonies. Krill grow in size until they die, so by their size we can measure their age. The proportion of juvenile krill in a sample tells researchers which years were good krill breeding years. As with penguins and seals, a krill's first year is a critical period for its survival into adulthood.

Female krill lay their eggs in offshore waters, where they begin to sink. The eggs hatch under pressure, over a mile beneath the surface. When the krill larvae hatch, they are only one millimeter long and must swim to the surface, which takes their nascent forms about two weeks. During those two weeks their mouths develop, enabling them to feed at the surface. The young krill survive their first winter by hiding under ice, squeezing into crevices and munching on the algae that grow on the surface. Ice is like a nursery for krill, where they are relatively safe from predators and have enough food to survive the winter. Once spring arrives and the sea ice melts, flushes of algal growth bloom in the fresh water left behind, and the krill have plenty of food. They grow into juveniles and no longer rely as much on sea ice, moving to the open ocean swarms that are so characteristic of this little shrimp. Krill reach maturity in about two to three years.

As adults, krill hang out in massive swarms in the top 220 yards of the water column. Out in open water they comb the sea with their feathery front legs and munch on the phytoplankton that they catch. Some adults in the krill population wander into deeper waters, foraging near the ocean bottom. Krill are an essential link between the seafloor and surface waters, shuttling nutrients between the two. Because krill swim constantly, they are also continually feeding and expelling waste. Krill waste, sinking to deeper waters, acts as fertilizer and also locks away carbon on the ocean floor. Every year, about 23 million tons of carbon dioxide are extracted from the atmosphere by antarctic krill, the equivalent of that produced yearly by 35 million cars.

Sorting masses of dead krill, I remembered a windy afternoon after Matt and I had done the nest census of the control colony, which was at the edge of a string of tide pools. I'd crouched down to peer into a tide pool by a rocky arch. The pool had been crusted with coralline algae, forming pink and green scabs on the sides. As I'd dipped my fingers in the clear ocean water, I'd spied a larger shape: a live antarctic krill, about the size of a dime, stranded from its marine environment and swimming in circles. The krill's bulging black eyes had stared at a world that doubtless seemed shrunken compared to the vastness of the ocean. It paddled with dozens of tiny legs, propelled by a clear tail shaped like a mermaid's. It had a nearly transparent, flushed-pink body, with spots of red along its spine and a large splotch of green behind its eyes, where it processed all the algae it ate.

The whole reason we were on Livingston Island was krill: the subject of international treaties, the basis of the antarctic ecosystem, the fuel for our study species, an engine of carbon capture, a key component of the world's marine food web, and also a little creature I had seen paddling around a small pond. I remembered the way it moved through the water, ever so elegantly, an individual separated from the teeming masses that loomed so large and amorphous in my mind. Alive, beautiful, and singular.

———

When I was in college, I worked in my biology professor's lab cataloging otoliths. They were mounted on cardboard slides and came from penguin diet samples collected in the South Shetland Islands. The otolith is the only true bone in a fish: a small, roundish one that forms part of the inner ear. Otoliths are the only part of a fish that remains in a penguin's stomach long after the fish has been digested. You can learn an astoundingly specific amount of information about a fish just from looking at the otolith left behind: species, age, sex, even whether the bone came from the left or right ear. Otoliths are an invaluable tool when trying to understand seabird diet and broader conditions at sea.

I'd crouch over the microscope from my lab bench, peering at minuscule ridges in the bone, trying to stretch my imagination toward what kind of world these bones came from and who the people were who collected them. I followed the trail the bones left for me, heading south, past every continent I've ever known, past the stormiest ocean passage in the world, to a pinprick of land at the edge of Antarctica. In the skua shack, I placed the tiny bones of a fish's inner ear, retrieved from the stomach of a penguin, into small round cardboard wells for another student to pore over.

As a student, I'd imagined a cold, wet island bursting with penguins, but I couldn't have possibly imagined that to acquire otoliths a lot of things have to go right. Inevitably, at other times things go very, very wrong.

I definitely couldn't have imagined the sinking feeling of tipping a bird and seeing blood come out of its bill, or the wrench of returning that bird to its colony, hoping it would be okay even though it seemed more stressed than most. I wasn't quite ready for a dead penguin when I found it the morning after our fourth diet-sampling round, but I knew exactly in which diet sample it had been. I had hoped the blood was nothing, but I was wrong.

If a bird died, you've not only killed that bird, who might be over ten years old, but also its chicks, who cannot survive without both parents. Sometimes this happens—the water bursts the stomach, something else goes wrong, we kill a penguin. It hadn't happened in years, but the eventuality is covered as part of the permit because sampling diets is messy and intrusive and death is always a possibility. Permits are issued by IACUC, the Institutional Animal Care and Use Committee, which oversees the protocols in any US-based research that involves animals. There are also Antarctic Conservation Act permits, issued by the National Science Foundation, which apply to all expeditions headed to Antarctica from the United States that impact wildlife.

I was alone at the colonies when I found the carcass. Matt was off checking on his skua nests. If a penguin is lost during sampling, protocol says we must necropsy the bird and remove its stomach contents, to not waste the opportunity to gather important data. I brought the dead penguin to the back of the skua shack and cut through the feathers under its chest with a knife, right under the chest muscles, taut and deep red. I removed the stomach from the chest cavity and emptied its contents into a bucket, rinsing it with rainwater to get all the otoliths out. It was hard to believe that just yesterday this penguin was healthy and very much alive. I put the stomach back into its cavity and felt the crushing weight of a life extinguished.

I walked down to the intertidal, placed the penguin on some rocks, and sat a distance away, waiting. It took only about thirty seconds for a giant petrel to swoop down and start to gorge on the fat and protein. Giant petrels, true to their name, are almost as big as albatross, with a similar beak made for tearing at flesh. They are the main scavengers in subantarctic (just outside the antarctic circle) and antarctic marine ecosystems. Scavengers tend to have big bodies because their food source is alternately either scarce or abundant, and they have to fit as much food as they can inside them when it's available.

Soon more petrels arrived, and they tugged the carcass around the rocks, tearing, ripping, and gulping it down. The biggest ones shoved the others off and prowled territorially around the carcass, wings half-spread, keen eyes ever watchful. I stayed until most of the penguin was gone. The petrels were hungry. Maybe the carcass scene was not good that day.

The giant petrels made me uneasy. Their eyes looked unhinged: bright yellow with a beady black iris, always scanning for flesh. Their heads were often bloodied from being shoved into the latest find. They were a silent presence among much noisier species. I usually came upon them in large groups, especially when food was nearby. They were aggressive and opportunistic, but they stayed far away from us humans, an impulse most other wildlife on the island had not developed from the lack of terrestrial predators. Perhaps my distaste for scavengers came from my fear of death. An all too real reminder that we are flesh and blood, that from the earth we came and to the earth we will return.

I sat on wet rocks and mourned the life of a penguin, the first dead one I'd seen. I knew the petrels, like all scavengers, played an important part in their ecosystem. They consumed a carcass before it rotted, picking bones clean, recycling the energy quickly back into living cells.

Isolation can amplify things that would have less prominence in more populated places. Everything feels more intense. Danger to a friend and coworker feels real and immediate. Death feels close and ever present. Darkness lurks just around the corner, a contrasting foil to equally intense moments of wonder. In the intensity of isolation unexpected connections are formed. Isolation itself requires a frame of reference—isolated from what? I was separated from friends and family, from urban amenities and the conveniences of modern society. But in a deeply interconnected world, we are rarely truly isolated. Connection only changes form. I found myself thrown into connection instead with the crew around me, with the little girl I used to be, with the simple reality of death and dead things—subsequently, with the vibrancy of aliveness, and with the rich texture of my own life among windswept hills.

A group of giant petrels often congregated on the beach below one of my colonies, legs tucked underneath them as they sat on the rocks. The second they spied me walking toward them, they began the evacuation—running on land and then on water with large, webbed feet, wings spread open, taking off. They were so much like albatross. Some were mottled brown and beige, some were pearly white, most were somewhere in between. When they all ran noiselessly along the water, wings out, catching air, in all their different shades of white and brown, I found them beautiful. As if they existed on the edge of the world of the living and the world of the dead, coming and going like ghosts.

10.

Mid-January

By January 15, peak summer, the Cape was crawling with life: fur seal puppies on the beaches, skua chicks running around the hills like shrunken ostriches, and penguin chicks steadily growing in the colonies. The gentoo chicks were over three weeks old, and Matt and I had completed most of our twenty-one-day chick weighs. The chinstrap chicks were close behind. Taking twenty-one-day chick weighs was one of the grossest data-collection duties we had. The chicks, already spotted with guano, shat everywhere out of fear when I picked them up. They then went into a bag covered in shit from other chicks I'd previously weighed. I tried to beat back the tide of penguin excrement by rinsing out the weight bags in the intertidal every day.

The chicks were about half the size of their parents now, like round and squishy butternut squashes, and more mobile. They began to imitate the crooning call of the adults, albeit hoarsely and squeakily, and wander from their nests, poking at their neighbors and clambering over rocks. Already some gentoo chicks had crèched—the older ones, at three and a half to four weeks—and stood in the colonies close together without their parents.

The fur seal puppies had also become more curious, exploring small ponds near the beaches and banding together in little puppy gangs, four to five puppies strong, lurching inland or into tidal pools. The puppies found ways to entertain themselves while they waited near their natal beaches for their mothers to return from foraging trips. Those whose mothers were onshore for a few brief days cuddled up to the familiar form, nursing hungrily.

Not all in a young puppy's life was feeding and play. The breeding season brought out Antarctica's most fearsome predator: the leopard seal. To hunt, leopard seals lurked in the waters just offshore, hoping to snag a penguin on its way back to the sea from the colonies, or a puppy that strayed from the beach into deeper waters. I often saw leopard seals hauled out on the beaches, napping peacefully and digesting their latest meal.

Leopard seals are the second-largest seal in the world, behind southern elephant seals. Leopard seal faces are unique: distinctively reptilian, with a huge, pointed mouth and slits for nostrils. They are incredibly powerful and muscular, with a long, thick neck and huge jaws. They are equipped with sharp, curved canines for ripping the flesh of large prey such as penguins and seals, and interlocked postcanines, which they use for eating krill and sieving plankton from the frigid waters. Leopard seals have a remarkably varied diet. They also hunt fish along the ocean floor. They measure between two and three yards long, lying smooth and sinewy on the rocks.

They are the closest thing I have seen to a sea monster—like a snake that's a lot bigger and stronger than you, aquatic, and carnivorous. If I encountered them in the water, I would shit myself and then die. But on land, they just seemed sleepy. The first few times I saw one hauled out, I watched, baffled, as penguins wandered past and fur seal puppies played nearby, while the carnivore, who would devour both creatures in the water, just napped on. There seemed to be some kind of truce on land, where marine animals could go about doing what they needed to do without worrying about hunting or being hunted.

The hunting grounds around the Cape were dominated by a few old matriarchs. Melba, one of the largest leopard seals in the area, was known and named by the crews that preceded mine. She was a particularly fearsome example. The first time I saw her, she was lounging on a beach by a group of gentoos. Melba seemed to have a sixth sense for the presence of people. Maybe this is why, as I peered at her that day, although her eyes seemed closed, she opened her mouth in a great yawn, rows of multiuse teeth glinting in the sun like a chest of weapons. Melba's maw was pink save for some dark spots on the roof of her mouth. The void she exposed seemed too big for the animal behind it. I stared, agog. Whether or not Melba meant it as a warning, it certainly felt like one. I walked slowly away, eyes wide, stepping on rocks more carefully than usual. I kept my distance and watched the penguins waddle by on the beach as Melba rode the smooth river of her slumber.

Leopard seals are fiercely competitive and often bear scars from battles over territory and food. These scars can identify individuals if they haven't yet been tagged. I realized the yawning seal was Melba when I saw pictures of her later. The seal crew had been trying to tag her flipper for years, but every time they snuck up while she was napping, she just happened to wake up, foiling their plans.

Besides the core antarctic fur seal monitoring, the seal team also took note of tagged leopard and elephant seals, particularly on Friday Cape-wide phocid surveys. Understanding the foraging habits of leopard seals has become critical to unraveling the ecological dynamics of this remote ecosystem. Leopard seal predation is thought to be one of the main reasons the Cape's population of antarctic fur seals is declining.

Leopard seals love ice, meaning they prefer to haul out and hunt from ice floes floating on the open ocean, eating fish, krill, and the pups of crabeater seals, another ice-loving species. Despite what their name so blatantly suggests, they do not eat crabs. Crabeater seals are uniquely adapted to hunt almost exclusively on antarctic krill, using their sieve-like tooth structure to filter them from water. Crabeaters live on the open ocean and have a circumpolar distribution. They haul out and mate on ice floes throughout the year; unlike fur seals, they do not form rookeries. As with leopard seals, female crabeaters simply hop onto the ice alone and have their pup. Males will attend to mother-pup pairs until the recent mothers are ready to mate, after which the males will depart. There are currently no particularly accurate estimates of crabeater seal populations, as their wide distribution in the open ocean and their solitary habits make them hard to study.

Since 1979, ice habitat in the Western Antarctic Peninsula has decreased by almost half. This has huge, rippling effects on the region. With less ice, leopard seals have resorted to hauling out on land. The leopard seals that show up at Cape Shirreff also take advantage of the abundant food sources there: the clueless puppies, fat with milk, and penguins zooming around in the water. This was not always the case—leopard seals prefer ice when it is available, and before 1996, no more than two leopard seals were ever seen foraging at the Cape at the same time. The number of leopard seals observed at Cape Shirreff rose sharply between 1998 and 2011. Since 2010, an average of 70 percent of the pups born have been consumed by leopard seals per year.

While at the beginning of the season we saw the occasional leopard seal every other week, in mid-December, weekly phocid surveys captured four, maybe five leopard seals hauled out. In the middle of January, the seal crew recorded eleven. By then we'd lost half of the thirty study puppies, similar rates to recent years, but it didn't make it easier to deal with. The fur seal females, desolate, howled and howled into the wind. The haunting call blows across the whole Cape, ricocheting against bare hillsides, an animal grief, raw, desperate, poignant. The Spanish name for fur seal is *lobo marino*, "marine wolf," and it never felt more appropriate to me than then.

After a foraging trip at sea, the fur seal females waited on the beaches for their offspring to show up. It wasn't that strange for a puppy to be absent from its home beach when its mother returned. The young fur seals that survived grew larger and more adventurous, and some wandered impressively far from their home beaches. The females called out to their puppies, bidding them to return. Usually after a female's absence, the pups were hungry and eager to return and gorge on thick, nutritious milk. Young fur seal puppies and their mothers can recognize the exact timbre of each other's calls. A puppy was officially counted as dead if it hadn't shown up within four days of its mother's return from foraging for food in the waters near the Cape. We didn't usually witness them getting eaten—the puppies simply disappeared.

Fur seal puppies that we had named and babysat were lost to the leopard seals every day. Cape Shirreff tradition demanded we toast each lost study puppy with a glass of Scotch ("Scotch guard"). Sometimes we had to toast a bunch at once—it was hard to keep up with the appetite of leopard seals. One particularly heartbreaking week, we lost little Narwhal, Cricket, and Dung Beetle. Matt reminisced on Cricket's namesake chirpy call and her slightly blond head. I remembered gentle Narwhal curling up in my lap. Dung Beetle was Sam's favorite, earning the name for being the fattest puppy we had all season.

The fur seal populations at the Cape are in a steep decline. Even so, antarctic fur seals are listed as a species of "least concern" by the International Union for Conservation of Nature (IUCN) because the global antarctic fur seal population is counted as one stock. The long-held narrative of the antarctic fur seal population in the South Shetlands is that after being nearly exterminated by sealers in the nineteenth century, seals from South Georgia recolonized the South Shetland Islands. This would mean that South Shetland seals can be traced back to South

Georgia's population, which today accounts for 97 percent of all antarctic fur seals.

Among the 3 percent that do not breed in South Georgia are those that breed in the South Shetland Islands, including at Cape Shirreff. With most of the population in South Georgia, the Cape's declines don't seem so critical—except that the Cape's populations are probably more different than previously thought. A paper Doug published in 2022 with decades of Cape Shirreff data argues that the antarctic fur seal populations in the South Shetland Islands are actually genetically distinct subpopulations. Instead of coming from South Georgia, small numbers of seals in the South Shetlands likely survived the sealing era and went on to recolonize their historic breeding beaches. In addition to genetic evidence for this split, seals at the Cape are physically larger, a common adaptation to a colder climate.

Subpopulations at the edge of a species' range are usually reserves of genetic diversity. A diverse gene pool is critical to a species' resilience and could play a key role in helping antarctic fur seals adapt to the changing environmental conditions brought about by climate change.

———————

The fur seal harems at the Cape were on edge. I didn't escape the vortex of maternal despair by virtue of being human. On a blustery day in January, I was walking across a beach to get to my penguin colonies as I did every day, and a fur seal female started growling and chasing me. This had happened all season: the seals would leap toward you menacingly but stop short of actually attacking. It's a warning, much more commonly displayed by subadult males. I would walk off and the females would go back to whatever they were doing before. But this female was not simply warning me. She followed me about fifty yards until I turned around and tried to get her to back off. I couldn't do my penguin rounds if I was constantly escaping from this creature. I poked her a few times with the ski pole I took everywhere for this purpose, strapping it to my pack on longer hikes. She didn't even seem to feel it. She was looking straight into my eyes, huffing. I feinted toward her at the same time she feinted toward me—the contact seemed to startle both of us, and she, committing, tore my pants at the knee. I thought, Fuck. She lunged again and that's when I started running.

I was hoping to lose her by virtue of my bipedal speed, but she sprinted right after me, throwing herself onto the rocks in a furious rage. I raced across slippery rocks caked in penguin shit, the forty-mile-per-hour wind tossing me around and making my eyes water. I cast a glance back at my assailant and caught her in hot pursuit, sharp canines bared. We bounded across the low spots between penguin colonies, errant chicks scattering into tight, panicked groups, fluffy noodle flippers flailing. I clambered up a steep ridge to the biggest chinstrap colony. Like a scared child hiding in her mother's skirts, I crouched low among the penguins. My pursuer, unwilling to climb a hill and confront belligerent chinstrap slapping, gave up the chase and glared after me in unresolved fury. Perhaps she'd lost her pup and decided I was the cause. Perhaps she was hurt, and angry. Perhaps she chased me because she couldn't chase a leopard seal.

———————

In mid-January, dark and heavy with its losses, another person joined us in camp—already occupied by the season-long crew of four plus Doug and Jesse, and Federico and Renato in the huts next door. I was not in a welcoming mood. The additional visitor, Anthony, was a NOAA administrator who would be in camp for two and a half weeks to do a site assessment for potential remodeling. Before his arrival, we all spent a couple of mornings putting up a temporary waterproof shelter called a WeatherPort, since all other beds were occupied. It was sturdier than a tent, framed with metal poles, but colder than anything with real walls. It blocked my view of the glacier as I left the outhouse in the morning but looked sort of lonesome and poetic off on its own, plunked on pallets and plywood on a little rise.

Anthony's journey was undertaken via an unlikely series of logistics: after flying from Chile to the Chilean base on King George Island, he helicoptered to a Chilean ship, rode a Zodiac to the Spanish base on the other side of Livingston Island, and was then taken by a clan of Spanish mountaineers across the glacier on a snowmobile. They had just been over our way and were happy to give him a lift. Apparently the Spaniards took a different route the second time because the previous route was too "crevassey" (as Jesse put it, "too death-y"), which did not do much to put Anthony at ease, especially because he was positioned in the front of the party, clutching a brightly clad mountaineering Spaniard who was ready to deliver him with a dramatic flourish.

After hiking home from the penguin colonies, I found out that I'd missed the Spaniards by mere minutes. "But they sent their regards," Doug assured me. They'd brought us a bottle of wine (Tempranillo) and a round of Spanish cheese (*queso vasco de oveja*), a couple of T-shirts from their base, and some stickers. The wine was wonderful, and suddenly feeling patriotic, I claimed a T-shirt as a consolation prize for not having been able to hang out with my countrymen.

Anthony was one more addition to a packed hut. While he slept in the WeatherPort, he ate and spent his days in main hut. In the morning while we bustled around getting ready for the day, Anthony would try to find somewhere out of the way to sit or stand, but every time he thought he was okay, he'd have to move so someone could get to something, shifting around the hut and pressing himself into ever-tighter corners. Trying to get a turn on the one email computer became increasingly tactical and difficult as all seven of us clamored for a moment to read emails and reply to our friends and family. At the dinner table, we squeezed in, elbows tucked against our sides.

In the mornings, Matt and I escaped as soon as we could to the skua shack, staying out as late as we could reasonably justify. My head was always in the penguin colonies, chicks now steadily crèching, one nest at a time, adults becoming scarcer and scarcer, as they no longer needed to attend to the chicks full-time. The crèched chicks started huddling together, about three-quarters the size of their parents, and spent their days chirping at one another and figuring out that climbing on rocks was awesome.

Matt and I were deploying geolocators (PTTs) and time-depth recorders (TDRs) on adults that were still guarding their chicks, and on adults with crèched chicks. We only had a limited number of the small devices, eleven in total, so we had to make sure to get them back after a week so they could be redeployed. Matt and I had deployed the devices on gentoos, recovered them a week later, deployed them on nine chinstrap adults, recovered them, and were preparing to deploy them again on gentoos whose chicks had crèched.

Once the little devices were recovered from the penguins, Matt and I cleaned them, downloaded the data, checked their programming, and electrical-taped them up. We attached the device to a penguin's back with small black zip ties and superglue. When I caught the bird a week later to remove the device, I cut off the zip ties and then cut the device out from the electrical tape that was superglued to the penguin's back, leaving behind a small strip of black tape that would soon fall off when the penguin molted. Matt and I spent our days at the colonies checking our nests for crèched chicks, weighing any that reached twenty-one days, programming and taping devices, deploying them, and then watching keenly for the penguins' return a week later.

With camp packed to bursting, we arrived back from the colonies later and later. We'd always taken turns making dinner during the season, and it was usually ready pretty late, so we'd push it—how late we could get back to camp leaving as little time as possible before dinner. At 8:00 or 8:30 p.m. we rolled in, using the excuse that we were waiting for a bird with a device to return to the colonies, loath to spend one more minute than necessary among the throngs.

We had been at the Cape three months, with two more to go. With little solitude or space to be introverted, Matt seemed to grow silent and withdrawn. He is generally a private person, and I could imagine how camp could have become his perfect nightmare. He would vent at the skua shack—I felt that he was stressed by our workload and couldn't find a way to unwind in peace. I'd gripe about food or the email computer, and we'd go on like aunties complaining over their knitting.

I often escaped to Little Chile in the evenings, which was wonderfully welcoming and spacious. Federico, Renato, and I sometimes watched movies in their main hut, or Renato and I hung out in the captain's hut, a smaller hut near the main Chilean hut, to have our own space. Renato and Federico would be around for only a couple more weeks.

I showed up one evening in the captain's hut, as previously agreed and as early as I could sneak away after dinner. Renato had brought a bottle of tequila and his speakers, running on downloaded music and charged by solar panels. He enjoyed hearing about the goings-on in camp gringo, and I entertained him with all the weird little dynamics of all these personalities crammed together: how Jesse was always too cold and Matt was too hot, leading to silent battles over how open the window should be; how Anthony was unaccustomed to camp life and particularly to the hygiene standards; how Whitney was always making us margaritas and how quickly they disappeared; and how Doug's stories grew more animated with every passing margarita.

With a passion for music, dance, and marine ecosystems, Renato was a ball of excitement and playfulness. I saw myself in the way his eyes lit up when he talked about something that fascinated him, off in his own world. For his master's he'd studied kelp forests in the Pribilof Islands, north of Alaska's Aleutian chain, where Matt and I first met. Renato's time near the Arctic awakened an affinity for polar places, and he looked for ways to do his PhD in Antarctica. He got his chance when a professor he knew needed a diver to collect data on kelp forests in the Antarctic Peninsula. While working and diving from Chile's base on King George Island, he met Mike, who was transiting from the Cape. Out of an initial conversation Renato developed his PhD project, in which he pivoted to terrestrial work with antarctic fur seals.

Renato asked me how I was doing, truly and wholly, and I collapsed into his support. Hanging out with him was like a soothing embrace, but it never felt serious. A burst of connection in a distant place, where we fit together amid rocks and snow and penguins.

———

With camp packed to bursting, the time I spent alone in the penguin colonies or hiking through the hills became precious and prolonged. There were moments of quiet, resplendent poetry. There were also things reined in the quiet, and resplendent, that made them mundane, human, accessible, such as my daily routine: breakfast, then suiting up for the trudge to the colonies, pulling on snow-resistant overalls, rubber boots, a fleece, and a windbreaker. Sunscreen, hat, sunglasses. Checking my pack: radio, field notebooks, pencils, lunch, snacks, layers, hand warmers, binoculars, water bottles. The walk to the colonies, usually alone—Matt tended to head out before I did. At the skua shack, changing into my penguin rubbers, and going through my colonies in the same order, checking on all fifteen plots, all thirty-eight known-age nests, then heading back to the shack. After lunch Matt and I carried out any deployments needed or any chores: shack maintenance, moving water barrels, cleaning, preparing or repairing gear. In the evenings, at main camp, we cleaned up data, wrote emails, read books, hung out with a drink in hand. Everyone in camp traded off dinner duties, which was my favorite domestic chore. I planned my menus for days, devising combinations and riffling through cookbooks for ideas. The nights when it was my turn to cook offered an opportunity to indulge my creative impulses and come back to myself.

For my family, adrift among new countries and new cultures, food has always been a grounding force. The staples my parents cooked connected me to the places they were from: spicy and chocolaty *mole poblano* from Mexico, my mother's specialty, or my father's Spanish paella or *pisto*, a stir-fry of summer vegetables. We mixed those dishes with food from the countries we lived in, combining and concocting a more abstract sense of home, of rootedness. My mother's cooking was a product of her amazing ability to create community around her wherever we went. Her recipes were picked up from other parents, from the veggie vendor, from friends made through proximity: South African bobotie, Egyptian falafel, sweet cookies and pies like steaming volcanoes. While my mother did most of the day-to-day cooking for my sibling and me, my father's rare day off started with a menu and ended with an elaborate meal, his entire day spent in the kitchen teaching us about knives and chopping and ingredients, an apron always tightly tied around his middle. Food was ceremony: it told the story of who we were and where we'd been. When I moved to the United States for college, my mother's parting gift was a cookbook of all our family recipes, like an anthology of all the cities and countries that we had been through. At the Cape, I summoned familiar dishes from well-worn handwritten pages and stale provisions and always felt comforted. I had added to the recipe book over the years, trading recipes with my field crews—I had lists of scrawled recipes for Thai food I'd learned to make on Midway Atoll: *laab*, *pad thai, tom yum kha.* A salad dressing a field crew lead had made on St. Lazaria, the pineapple upside-down cake my coworker and I had made on St. George for Matt's birthday. I added the islands I worked on to the anthology, charting my life in food.

At the Cape, food was one of our few luxuries, and vegetables in particular were precious treasures. One night, Anthony cooked all the remaining eggplants, which were (and remain) my favorite vegetable, but didn't do it right—they were rubbery and undercooked, and then they were no more. "What a waste!" I fumed in the skua shack the next day, while Matt nodded sympathetically. "And now we have no more *fucking* eggplant." The anger and indignation boiled in my chest. My life was condensed to a pinprick—at no point did I scrutinize my rage, although now it seems absurd.

In what we call the real world, time is categorized: work, leisure, play. People are categorized: family, friends, coworkers, employers, strangers. Spaces are categorized: office, kitchen, bedroom, living room. The person you are at work is probably different from the person you are with your friends or alone at home. Out there, all of those categories fell apart. We had one single room, one group of people, one way of life. Did it count as work when it was your turn to cook or to dump the slop and pee buckets?

This immersion could mean many things. Sometimes I felt as if I were living at the office, or that I lived at summer camp or at a sleepover. As if I never worked; as if everything were work. There are benefits to this breakdown of official boundaries; I had a social opening that meant I got to hang out with my bosses as friends, get to know my coworkers better, and relax into the crew. Lead researchers such as Doug and Mike have a lot of power in these situations, where they are responsible for a small, isolated group of people. With Mike, I was immersed in history, on a roller coaster of previous years and decades of field camp stories. With Doug, everything was about the future: how we could improve camp, what would be changing, what studies we'd be working on.

The researchers I worked with at Cape Shirreff were wonderful. They made me feel safe and supported, and I always enjoyed the time I spent with them. I'm lucky, though; where power concentrates, it can be abused. In 2014, 666 scientists from across disciplines were surveyed on their experiences in the field. Field sites are usually active research areas where a team goes to collect data away from their usual lab or research center, where researchers and their crews or students stay together in field housing. Of those surveyed, 64 percent had experienced sexual harassment and 20 percent had been sexually assaulted by their lead researchers or other members of the crew. These numbers are devastating. I've heard many stories about techs and grad students fending off the unwelcome advances of their coworkers or superiors. In such a remote setting, dealing with sexual harassment completely transforms the experience. I know I wouldn't have enjoyed being in the field as much as I did or gotten as much out of it if I had constantly been looking over my shoulder, avoiding someone, feeling threatened. I cannot imagine. The field can be uplifting, magical, and inspiring, but it must first feel safe.

11.

Late January

In the Heroic Age of antarctic exploration, pre-1922, European nations, as well as Japan and Australia, sent men to brave the continent's harsh conditions to map, chart, and document new land, building the case for territorial claims. Explorers scrambled for funding from private investors, promising profits from expedition books and lecture series. These expeditions, seventeen in all between 1899 and 1922, sought to investigate the continent's vast landmass. As with Mawson's expedition in 1911, ships dropped parties off at the tail end of the austral summer (January or February) when the sea ice was broken up and ships could better access land. Expeditions built huts onshore and holed up for the winter, preparing to set out as soon as spring broke and the weather improved.

Two years before Mawson's harrowing journey, the geographic south pole had been reached by an expedition led by Norwegian explorer Roald Amundsen. Amundsen is undoubtedly my favorite antarctic explorer, and after reading a biography of him, *The Last Viking*, in the years that followed my time at the Cape, I developed a bit of a crush on him. (How could I not, with a title like that?) He is often called the greatest explorer that ever lived; he was also the first to cross the Northwest Passage, and the first to have been to both poles.

Amundsen was singularly focused on his goals and pursued them with fierce intensity. A meticulous planner, he prepared every aspect of his expeditions with care, thinking everything through ("Adventure is just bad planning," he once wrote). He was notoriously bad at building fanfare. Already close to polar lands, Norway's explorations had largely been focused on the Arctic.

One of the biggest challenges for arctic explorers in the late 1800s was finding a maritime passage along the north coast of the American continent that allowed ships to travel from the Atlantic to the Pacific Oceans. This route, if found, could completely reshape global shipping patterns by offering a much more direct route from Europe to Asia, faster than the current route around the southern horn of South America. After centuries of attempts, Amundsen was the first to successfully cross the Northwest Passage by boat on a journey that took him three years, from 1903 to 1906. The ship was stuck for two years in pack ice near an island north of Canada. He settled in the area with his small crew of six, living with and learning from local Inuit communities. His ship finally broke free, and Amundsen and his crew completed the crossing, landing in Alaska. At which point he skied five hundred miles to Eagle, Alaska, where he could send a telegram announcing his success, did that, and promptly skied the five hundred miles back to his ship.

Ruminating over his motivations while hiking to the skua shack, I often wondered what it was about the poles that held such magnetic attraction for Amundsen: Why, when he returned from one expedition, did he lose no time in planning the next? I wondered how the quality of his inner life changed when on the ice. His biography has excerpts of his writing, which betrayed little of the man beyond what was publicly revealed—he was not known for being effusive. Whatever truths he discovered he kept to himself.

Upon his return to Norway from Alaska, Amundsen set his sights on the north pole. In 1909, in the middle of preparations, he received reports that two Americans—Frederick Albert Cook and Robert Peary—both claimed to have been the first to reach the north pole. The north taken, he shifted his sights to the south pole, though he continued his preparations as if he were still planning to head north. The funding for Amundsen's expedition—granted by the king of Norway and the Norwegian parliament—was based on his initial plan to head deep into the Arctic. Amundsen did not want to risk losing funding by announcing a sudden change in plans. Amundsen also knew that the Englishman Robert Falcon Scott was organizing an expedition with the same goal of reaching the south pole, furthering the need for secrecy. The prospect of being the first human to ever step foot on the south pole was a huge, career-defining prize for both Amundsen and Scott.

Amundsen took care to select his crew of nineteen, which included naval lieutenants, a ski champion, a Russian oceanographer, a Swedish engineer, plus a few men who'd traveled with Amundsen through the Northwest Passage, including the cook.

The *Fram*, Amundsen's expedition ship, sailed from Norway in August 1910. Once Amundsen and his crew were well underway, he informed his crew of the plan change and asked them individually if they would still join him. Once they recovered from their shock, all agreed to join his antarctic expedition. From Madeira, Amundsen sent Scott a telegram: "Beg to inform you *Fram* proceeding Antarctic—Amundsen."

Four months later, in January, the middle of the polar summer, the *Fram* set anchor in the Bay of Whales, on one side of the Ross Ice Shelf, in East Antarctica. The expedition needed enough fair weather to set up their camp on land and lay depots. After a month of this, the crew settled in for the winter. Amundsen enforced a strict schedule to keep the crew sane in the long dark night. They fine-tuned equipment and planed down sledges to reduce their weight. They sewed new, lighter tents from the materials they'd brought with them on the ship. Amundsen's intention was to ski and sled their way to the pole as soon as the weather warmed. They had with them ninety-seven dogs from North Greenland, the most cold-hardy breeds Amundsen could find. The crew also took scientific observations, measuring wind strength, temperatures, pressure. Both Scott and Amundsen intended their expeditions to be both the first to the pole and also scientific missions, collecting observations and samples of the landscape they traversed.

Scott had set up his base camp on the other end of the Ross Ice Shelf, some three hundred miles west of Amundsen's camp. They would be attempting the pole at the same time, crossing the same ice shelf toward the heart of the continent. On its way to Ross Island, Scott's ship had encountered the *Fram*, anchored in the Bay of Whales. Lunches were hosted on each ship for officers from the other expedition, and the parties went their separate ways civilly.

After one start in September, thwarted by a blizzard and terrible cold, Amundsen's south pole team (five men, four sledges, and fifty-two dogs) headed out on October 20, 1911, struggling forward through strong winds and their fair share of blizzards. To navigate, they used a sextant, which measures the angle between the horizon and a celestial body—in this case, the sun. With an accurate clock, and some calculations, early explorers could determine their position in latitude/longitude coordinates with impressive accuracy. Every day when the men broke for lunch, they built a cairn on the flat glacier to aid in their return. Some of the dogs were killed to conserve the rations needed to keep a crucial number of dogs alive. On December 14, Amundsen and four others finally reached the south pole, almost two months after they'd set out from their main base at the Bay of Whales. There was no sign of Scott: Amundsen had won the race. The return journey was as smooth and efficient as it could be under the conditions, and all returned safely.

Scott's expedition was not as triumphant. Scott's crew included five veterans of previous antarctic journeys, many navy officers, and a small group of scientists headed by Edward Wilson, a medical doctor and zoologist. Scott had brought ponies, dogs, and motorized sledges to aid his team across the rough terrain of Antarctica. The ponies and motors did not last long, and the dogs were mismanaged. The crew themselves were forced to pull heavy sledges through the snow. Their progress maddeningly slow and arduous, Scott and his men made a fraction of the progress of Amundsen's team. They also carried the impossible expectations of a great imperial power and trusted it entirely: the British way was the best way. Scott relied on technology used or developed in England. Amundsen, by contrast, had spent years with the Inuit, a people accustomed to the extreme conditions of the poles, during his three-year crossing of the Northwest Passage. Amundsen learned critical skills that he applied to his antarctic expedition: dog-sledding techniques and technology and Inuit designs for his men's clothing, incorporating wolf and reindeer skins. Tellingly, the Inuit felt no need to be the first to any pole, although their technology was indispensable to those who did.

Scott's expedition reached the pole after a long and difficult journey, one month after Amundsen. Tracks, camp remains, and a Norwegian flag made it clear that they had lost the race. They began the struggle back but were not able to reach a food cache in time. Every member of Scott's party died on the return trip. Their remains were found by later expeditions, and Scott's diary was recovered, which narrates in harrowing detail the crew's struggle to survive as they labored toward food depots through constant injuries and storms, eventually succumbing to the elements.

Amundsen found out about Scott's ill-fated expedition after his return to Europe. The British tragedy, in some ways, overshadowed Amundsen's accomplishment. It's hard to claim victory over a martyr. The US base on the south pole is named after both men: the Amundsen-Scott base.

As the story is often told in contemporary times, Scott was a bit of a narcissist, obsessed with his own heroism. Roland Huntford's 1979 account of Amundsen's and Scott's expeditions, *The Last Place on Earth*, is filled with disdain for Scott's leadership abilities and judgment, though Scott's defenders claim he was beset with unusually bad weather and bad luck. While I'm tempted to believe Scott had a bit of an inflated ego—especially compared to Amundsen's controlled, understated competence—Scott still endured more than I could ever imagine, and his ambition is worthy of, if not admiration, then at least respect.

Lofty words for one guilty of having fun at Scott's expense. I first heard about Amundsen's and Scott's expeditions from Matt—who had read about both expeditions before I did—in the skua shack. We couldn't help but poke fun at Scott's dramatic, Victorian heroism.

"It will be ours! Because we are British! We will arrive, and we will win!" I declared, looking nobly into the distance.

"We will land on the continent, and we will have it!" Matt proclaimed, hitting the table with his fist. "For the queen!"

Snorting, I took another sip of my tea, more than a little grateful for the wonders of propane and plywood, which had replaced paraffin stoves and canvas tents.

Over time, dramatic stories of the Heroic Age have been so mythologized as to serve as a creation myth for the human presence in Antarctica, Amundsen's and Scott's stories in particular. These narratives represent an upper-class European perspective on what it means to encounter and travel through new land: competition, nationalist glory, the search for exploitable resources, and the individualism that so often fuels heroism. The history of antarctic exploration is often told from this perspective, too, and decontextualized from the history of profit and colonization that enabled Europe to afford expeditions in the first place. As if the accomplishments of explorers were individual feats and not the machinations of world history manifesting on a continent yet unknown. The same frameworks of exploitation (sealing, whaling) that made England rich recurred on a continent that happened to have no Indigenous population.

In *Class and Colonialism in Antarctic Exploration*, Ben Maddison describes how European explorers in Antarctica faltered against a landmass that had no Indigenous communities to dispossess—the usual patterns of colonialism absent, many "satisfied their colonial needs by portraying penguins as surrogate native populations." French officers, in claiming Adélie Land, in East Antarctica, for France, described how French sailors moved up a rocky slope to plant a flag and "hurled down the penguins, who were much astonished to find themselves so brutally dispossessed of the island." The British captain James Clark Ross, in claiming Victoria Land, described penguin opposition as "pecking us with their sharp beaks, disputing possession." On Cook's expeditions, a midshipman described the penguins as the "Islanders," a term used by colonists in reference to Indigenous communities on Pacific islands. He went on to describe how crew movements on an antarctic island were "completely arrested by these Gentlemen," penguins, who "disputed our right to proceed."

I often wondered what early narratives of Antarctica would look like outside the paradigms of competition, colonization, and profit. Beyond territorial claims and nationalistic fervor. I wondered what different societies would have made of the ice and snow had they reached it first, and how they would have incorporated it into their own culture and mythology. I turned over these questions in my mind far before I learned of Hui Te Rangiora, the Māori voyager who reached polar lands.

In New Zealand, a research project run by the country's Ross Sea Region Research and Monitoring Programme explores similar questions in the context of *te ao* Māori, or a Māori worldview. The project is called "Māori in Antarctica: Ka mua, ka muri." *Ka mua, ka muri* means "walking backward into the future," a Māori proverb that refers to the practice of drawing deeply from ancestral knowledge to guide decisions that have a bearing on the future. Through carvings, seminars, and workshops, the project works to understand the ways Māori knowledge and ways of knowing unfold in Antarctica. Topics considered include the importance of the *te ao wairua*, the spiritual realm, as well as science when engaging with the continent, the difficulties of working within a state-centric governance system, and the rich Māori history with the Southern Ocean. In one seminar, Dr. Krushil Watene reframed the continent as a teacher that pushes us beyond our usual frameworks, as a subject to be learned from, rather than an object to be explored. What, she challenged, does Antarctica teach us about the limits of our current thinking?

In the penguin colonies, all my gentoo chicks and about half of my chinstraps had crèched. I stopped checking my plot nests and known-age nests after crèche because the whole nest structure fell apart, chicks moving around and huddling together. We did not keep track of the study chicks after that—if a chick made it to crèching, the nest was considered to have been successful. Matt and I were busy with our crèche deployments of PTTs and TDRs and conducting the fifth and last rounds of diet sampling on both chinstraps and gentoos. On January 24, we conducted the gentoo chick census by each heading to our designated colonies and counting the chicks we saw per colony—a total of 902 chicks. All the skua chicks had hatched and were growing fast. Eighteen of the thirty original study fur seal puppies had died, likely leopard seal kills.

A day before my twenty-fifth birthday, at the end of January, Renato and Federico, after two months in camp, got picked up by a Chilean helicopter from King George Island. The helicopter also picked up Anthony after his two and a half weeks. We knew they'd be leaving soon, but helicopter arrivals could be abrupt and opportunistic, launched when the weather was calm. I heard about it on the radio at the skua shack, and I changed and immediately hiked back to main camp after my rounds to attempt a proper goodbye, instead of lingering at the skua shack with a mug of tea, waiting for device-toting penguins to return. I made it, barely, and got a few hugs in before the three were whisked away on a giant metal bird. I was sad to see my friends and partners in occasional debauchery fly away. I would miss Renato, and I'd felt that in Little Chile I could manifest the part of me that had grown up in Chile and Argentina, that tended toward loud proclamations and dramatics, with all the exuberance Latin American culture demanded.

Sometimes I felt as if I lived in a snow globe. My world was small, with rigid social and geographical edges. It was cold, it snowed. The sky was huge and concave. The rhythm of my life was settled until someone picked it up and shook it, throwing everything into chaos again. When the world within the glass settled once more, everything ended up shifted around.

After being so tightly packed, the crew exploded outward like a spring. Whitney moved out and took over the WeatherPort, where Anthony had slept. Matt, traumatized by the months-long claustrophobia, picked up and moved into the newly vacated Chilean hut. Over the seasons each country's visiting researchers kept to their own huts, and we were unsure what precisely was allowed, but no one would be coming to the island to check that everyone was in the right place. The whole crew still largely lived and ate in main hut in the evenings, but everyone dispersed after dinner to their respective hovels.

At night, with Matt in Little Chile, Doug and Jesse holed up in the "old fart's room," and Whitney off in the WeatherPort, Sam and I wandered around main hut, feeling like the king and queen of our little four-walled empire. Left to our own devices, Sam and I began crafting a rubber-band ball from a bag of rubber bands that had been floating around, diligently adding the bands to the ball one by one, watching our creation grow big enough to be thrown into the small basket mounted above the door. We both tended to start brushing our teeth and then do other things at the same time, ending up absorbed in some task with a toothbrush dangling from our mouth, dribbling toothpaste. We'd get into a conversation and blabber on, garbling words and spitting foam.

Sometimes we talked about the small town where we'd both gone to college, about an hour east of Los Angeles. His school had a technical, computer-science focus and a reputation for being chock-full of brilliant nerds who hid in underground tunnels staring at computer screens. Equipped with chemical knowledge and fueled by the faulty reasoning of young adulthood, students concocted machines and toxic chemicals in their dorms, the campus a breeding ground for reckless genius. The walls of the school were fireproof for a reason. These were the stories we heard, at least, from the outside.

The liberal arts school I went to was known as the more pretentious campus, bursting with nascent intellectuals debating socialism, communism, the revolution, naive in a liberal academic bubble. I teased Sam for being a nerd, and he teased me for finding metaphors in everything, for having so many opinions, for always asking, What does it mean, what does it mean, what does it mean?

Sam's mind was far more concrete—in the field he was always asking, How can we make this process easier? More efficient? More effective? He'd caught me doing some Excel calculations manually when we'd been testing our radio tags and was baffled, telling me he could easily write a code for that and save me the headache.

The labyrinth of seal data was his biggest nemesis because it was nearly impossible to pull up consolidated tables of summarized information or to find specific information quickly. He stayed up late tapping away on his computer, writing code to smooth things out, untangling decades of ad hoc database use one knot at a time. Every monitoring program I've ever worked for was plagued with constant data-management headaches. To say computer-programming and data-management expertise were in demand would be a vast understatement. With field skills and technical skills, he knew what the needs were from both ends: how the data was gathered and how it needed to be analyzed. I could already see that he would always have work building digital bridges to connect the two.

If I were to be so lucky. I'm not as drawn to the analytical, quantitative element of research. Instead, I thought about what the Antarctic

Treaty system can teach us about solving global problems. I looked into the stark landscape and thought about the way the human psyche interprets and reinterprets land and place. My mind drifted in political and metaphorical eddies, drawn to the softer sciences along with my penchant for biology—sociology, psychology, politics, anthropology, art. I didn't feel like I slotted into place the way Sam did. At least, my interests didn't seem as materially necessary.

As field techs, our role models are usually the people that head the program, in this case the Antarctic Ecosystem Research Division, run by NOAA lead researchers such as Jefferson, Mike, and Doug. While most research-program leads have fieldwork experience, only a minority of people who do fieldwork go on to get master's and PhDs and end up as actual researchers and program leads. Being a research lead is a difficult trajectory—the PhD component alone is significant enough to dissuade many field techs. The job of a researcher is complicated and requires many skills: diplomacy, data science, fieldwork, scientific writing, logistics, the interpersonal skills to lead field techs on a long and remote season, and the flexibility to be absent from your "real life" for a couple of months every year. Not everyone is cut out for it, and not everyone wants to be.

Many other roles orbit researchers, facilitating their work in a million different ways: science communicators, logistics support, data managers, data scientists, graphic designers, program managers. People that help make the science come alive—teach it, apply it, enable it, interpret it. Sam would fit right in on the data-management side—there were already data-specific roles in San Diego within the program. Whitney would use the science in her veterinary practice. Maybe somewhere in this system there'd be a place for me, a person who loves to launch into philosophical musings and who needs ecology to understand herself.

Beneath the veneer of idle musing swirled all the pressures I felt to find the thing that would be my life's work and pour myself into it: anxiety, from a childhood overachiever, measuring my worth in proportion to the list of accomplishments under my name, a deep inner drive to protect the things that I found most meaningful in this world, and the cataclysmic specter of climate change, hovering like a shadow over my future, demanding with unmatched urgency that I use my strengths to make the biggest difference possible.

In Matt's self-exile, I spent more time with the rest of the crew, falling into their easy chatter, a penguin infiltrator in the seal circle. Matt came and went like a shadow, slowly moving all his stuff to the Chilean hut. Doug and Whitney seemed worried by what they saw as Matt's silence and his absence. I knew Matt was an introvert and liked his alone time, but he generally did well in the field, from what I'd seen. After the sardine compression of seven people in camp, I felt that he was tired of having people around all the time, needed space and quiet, and didn't want to have any conversations he didn't absolutely need to have. He told me as much, mentioning he would be okay not talking to anyone for the next two months (though I didn't count as a "person"—an introvert's highest praise).

When I saw Matt approaching main camp from Little Chile in the evening to check his email or eat, I'd warn the others not to make eye contact. Although he hadn't told me to, I asked the crew to pretend he didn't exist. Doug, Jesse, Whitney, and Sam would go about their business, dutifully ignoring him at my behest, facing away from him if at all possible. Sometimes the best way to help someone was to leave the person the hell alone.

The weather was steadily getting colder, dipping back under freezing, the wind relentless, the sky spitting down a combination of rain and sleet. More often than not, the return to main hut required powering through the wind—especially on my ongoing skua rounds on half the Cape's hills, marching up and down, up and down. I remember vividly the way those long, solitary battles against the wind felt in my body, the way my muscles strained against invisible energy, gusts blowing forty, fifty miles per hour. I felt like a giant battling prehistoric forces, before people, before animals, before life, as if I were stuck in a creation myth. After making it home I'd feel both whole and hollow from the effort, like a glass ball washed up after a storm.

I was tired from the rush of work of the past weeks, but still had my daily hike and daily nest checks. Our data collection was ending, and all the fun parts—catching penguins, deploying devices, banding—was almost completed. The weight of the long days, the work, the daily rhythm of the hikes, and the isolation all began to settle under my skin when there was little excitement left in the season. My brain slipped into a weary fog.

Some mornings, cocooned in my bunk, listening to the wind howling outside, I dreaded the world on the other side of the main hut door. I knew it would be cold, harsh, and wet. Later, as I trudged up a ridge—wind whipping me in the face, hail stinging my skin—all I'd want in the whole world was to curl up on a soft couch against someone who was warm and clean and who would maybe even fed me cookies while I watched goofy movies and slipped in and out of sleep, with all the peace of a blank to-do list. Why was I doing this to myself? What was I looking for, exactly, and how many more islands until I found it?

Life at the Cape could feel monotonous, as spectacular as our landscape could be. The days slipped by and blended into one another. I became even more accustomed to discomfort—such as my index knuckles being red and swollen from the constant cold. (I learned that this was called chilblain, and like poison oak it only gets worse over time.) Taking turns to put on sunscreen in the small mirror in main hut. Eggs on rice. Eggs on salmon. Eggs on wet, thawed spinach. Socks so stiff with use there was a right sock and a left sock. Mold in my water bottle. Mold on the walls. The feeling of shoving my toes into damp boots. The musty smell of weeks of accumulated sweat on the inside of fleece pants, fleece sweaters, fleece blankets. Clammy clothes that only felt dry after I'd worn them for an hour, warmed and dried by my body heat. Mud on everything.

Matt and I still headed to the colonies every day to catch chinstrap penguins from the last weeklong TDR and PTT deployment and check on our nests. We were also still going on our skua rounds every fourth day, looking for the skua chicks, who scampered and then froze still as rocks near their nests, in perfect camouflage. The chicks were safely hidden, unlike the adults, who dive-bombed my head, a sure indication that their precious chicks were, in fact, around there somewhere.

After the gentoo chicks crèched—big enough to be safe from skuas and to stay warm without a toasty brood patch to curl up against—the gentoo adults came to shore only to feed them. Parents decided when it was time for their chicks to crèche, but usually it was about four weeks from hatch. Since the gentoo colonies had laid eggs earlier than the chinstraps, the gentoo chicks were a week ahead in their development, and they started crèching sooner. About half of the chinstrap nests had crèched, while the other half still had parents guarding their chicks.

The gentoo chicks were almost as big as their parents but still covered in woolly down. While the adults are out at sea, the chicks stand around in the vicinity of their colonies in a tight group, huddling together when it's cold and lying splayed on the rocks on their bellies, woolly flippers out, when it's warm. When a parent was onshore, the chicks bent down and pointed their head up to catch the regurgitated krill.

The adult penguins returned from their foraging trips only to be chased around by a bunch of hungry, uncoordinated chicks. After running for a bit, the adults would turn around suddenly and nip at the chicks, as if saying, *Stop chasing me! I will not feed you!* The chicks bowed their heads sheepishly, noodle flippers dragging on the mucky ground. As soon as the adult's back was turned, the chicks renewed their chase and the charade continued. Some gentoo chicks in particularly muddy colonies had balls of dry mud at the end of their droopy, downy flippers and they swung them around like maces.

The flippers of the chicks lacked the rigid tone of a seasoned swimmer. Even after growing in their adult feathers, the chicks' weight was centered in their stomach, with the chest underdeveloped, body still in full digestion and growth mode. The weight of an adult is centered in the chest, where their powerful swimming muscles are, and their bodies taper down from there to their feet, forming a slim profile to complement the sharp lines of their tuxedos. So handsome, these penguins.

Our globalized world has come to know the continent largely through science—a discipline known as distant, cold, and impartial. A fitting framework through which to interpret a continent also seen as distant, cold, and impartial. But those of us who work with wildlife come to know their study species well and even grow close with an environment that promises no warmth or comfort.

Instead of the ultimate nonhuman landscape, I saw my humanness everywhere reflected back, or, rather, I saw that these things I knew as human were just universal laws for organizing matter, that I was so much more like moss or krill or penguin than I could have imagined, that the boundary between my legs and the dark, wet soil was thin, and largely imaginary.

During the day, I stood on the sidelines of the lives unfolding on the island. In my waterproof work notebook, I wrote the number of chicks I saw, the number of times a brown skua tried to snatch a young chick, the date a chick disappeared or the date I found one dead in the nest. In the evenings, I'd go back to main hut, crawl into my bunk, and write about how all of this made me feel. I tried to paint the picture of an expressive, emotive reality and send long emails to my friends and family. I did not write about the reproductive successes of the colonies or the results of our population counts. Instead, I described the feeling of a chick peeping inside the egg in my palm, the sound of an egg hatching, the sight of a newborn chick too weak to hold up its head, the tenderness of a parent, the wild cry of a fur seal. I tried to communicate how Antarctica, beyond the charts and maps of climate change, is, like any other continent, a place of grief, sorrow, joy, love, and survival.

12.

Early February

After most of the nests had crèched, Matt and I traded off conducting beach surveys, sweeping the shore below the colonies, looking for nonbreeding banded birds that might be coming back to the colonies for the first time. By checking for those known-age birds banded in previous years, we can know which penguins survived the first year of their lives, figured out how to swim and forage and be a penguin. Penguins start raising their own chicks usually between three and six years old. "Juvenile" penguins were no longer dependent on their parents but had not yet reached breeding age.

I was always reluctant to head out from the skua shack to do beach sweep, as it was the last thing I did in the day before heading back to camp. In February, we all stared into space a lot. Just staring at a wall. I could space out forever. Especially if I was in a room alone, with no one to hold me accountable for my time or to direct my attention. Sometimes at the skua shack, after Matt had gone back to main hut and it was my turn to do beach sweep, I had to work hard to rally, get up, change, go off on beach sweep, and hike home. The effort seemed superhuman. Once I sat down, it was all over. Even after my feet went numb and sounds were distant ambience and the light started to fade. Fifteen minutes, half an hour, forty-five minutes later, I would sit, staring at nothing. Wow. Floor. Look at that.

Once I rallied, I picked my way along the slippery rocks of the intertidal, hungry and tired, my mind everywhere but on the beach I was walking. I would see about one or two new known-age birds on beach sweep a week. In the thick fog, rocks and penguins emerged from nothing and disappeared into nothing. I'd be rubbed raw and cracked open by the daily grind, and these fissures exposed the vacuum in my heart to a rush of circumpolar air.

There were moments that snapped me to attention: a penguin stood in front of me, on a dark rock, pausing from its preening to look at me suspiciously. I read the number on its band with my binoculars before it waddled away into the mist, sleek and flushed from the ocean. It wore a precious known-age band, meaning it had been banded when it was a chick. With the quick-reference table in my notebook, I calculated that the penguin was thirteen years old. That penguin was born when I was eleven and fledged on these beaches in the months afterward. By chance, it was one of the 10 percent of chicks that get banded every year, and every year since then it had survived by hunting krill and returned to these beaches in the summer. In all that time, the band had not fallen off. There was the penguin, popping out of the water on the beach I was walking, at the time I was walking it, and by the confluence of a thousand little miracles we encountered each other and I could have this knowledge: that bird was thirteen years old.

I spent a great deal of time during beach sweep lingering at what I thought of as the chinnie train station. It was a great spot under the colonies—a few ledges of rocks leading into the intertidal and out to the ocean. This was where the chinnies came and went from the water. The adults were still returning to the colonies to feed their crèched chicks. The outgoing penguins were filthy, streaked with penguin pudding, exhausted from cramped colony life and hungry chicks, their feathers all out of whack. They waddled down to the rocks and waited for a critical number of penguins to assemble, all in a similar state. Once there was a group of them, they tentatively tiptoed up to the waves, and when a big wave splashed their toes, maybe one lurched into it, and others followed, leaping into the surf and disappearing entirely. Penguins use their feet and tail as a rudder and propel themselves through the water with their flippers, which are covered in feathers so tiny and tightly packed they feel like scales.

The incoming chinnies, returned from their foraging trips, peeked their heads out of the water upon approach, calculating the landing. I imagined they were making judgment calls and measurements, eyeing the rocks and the conditions, deciding upon a strategy, but no matter what they did, their arrival always seemed chaotic. Penguins tumbled in on the crashing waves, foam and feathers one whirl of energy, the birds scrambling onto the rocks and gripping on to impossible ledges with a single toenail. On calm days, or at low tide, they "squirted" out of the water (Matt's perfect description) and landed on their bellies, quickly righting themselves and looking around with as much dignity as they could muster. Which was a lot: incoming penguins were sleek, feathers smoothed by the water, tuxedos impeccably clean and glossy. Their feet and the insides of their flippers were pink, and they bristled with vitality, full of krill to feed their chicks. I love few things in this world more than a chinstrap penguin fresh from the sea.

After they emerged, they hung out on the rocks preening, putting all their feathers back into place and re-waterproofing them. Seabird feathers are waterproof by virtue of the uropygial gland, which sits at the base of the bird's tail and soaks a small cluster of short feathers with waterproofing oil. Most birds have the gland for feather maintenance, picking up oil from the gland with their bill and smoothing it over their feathers. As penguins emerged from the surf, beads of water slid down their waterproof feathers, and they shook themselves to fling the rest of the water off.

After a thorough smoothing and waterproofing session, I imagined they dreaded plunging into the filthy colonies and instead lingered, procrastinating. At beaches that were not as busy as the ones right below the colonies, the chinnies called before landing and called again on land, looking for other chinnies that might be around. The solitary wet chinnie call was hesitant and searching. When I heard it, I always stopped to find the lonely bird that emitted it: "Guys? . . . Guys?" it seemed to say. "Hey, guys? . . . Anyone out there?" They called until they found one another and ventured forth to the colonies in a tight little pack.

When the water was still and the tide was low, the stillness in the tide pools reflected the penguins' image, framed by pearls of algae strung across the rocks. I watched them picking their way across the slippery intertidal, flippers aloft, eyes calculating, passing by pools of their reflection like pearl-laced portals into another dimension. Right at the point at which the feet of a chinstrap penguin met the feet of its perfect mirror image, I and all my constructs disappeared.

———————

On February 8, Matt and I conducted the chinstrap chick census, two weeks after peak chinstrap crèche. It sounds orderly and straightforward, except that compared to the gentoo colonies, chinstrap colonies were a lot more numerous and were sprawled across whole ridges. The census required the both of us due to the sheer volume of chicks. For the bigger colonies, we had to section off the chicks in subsections. The biggest of the twenty colonies, up on a ridge and exclusively chinstrap (one of ten colonies that were chinnie-only), took us hours and seventeen subdivisions. To herd them, we walked a perimeter around a subdivision, the chicks scurrying off on either side of us to the colony's center. I felt like Moses parting the Red Sea, except it was fluffy and waddled. We could then count that section, but we had to keep an eye on the borders, because chinstraps never cooperate, and sometimes the chicks defected or decided they didn't like the group they were put in or just wanted to wander off. We each counted the chicks twice, and if we had three counts between us that were within 5 percent, we took the average of the three numbers and moved on. If we were more than 5 percent off, we counted again until we got it right. To make it more complicated, some adults were in the mix that were not pleased about this mass chick movement because of the crowding, so they bit and slapped to scatter them. I couldn't move too quickly because then I'd cause a stampede and mess up all the groups. It was a long day. We counted 3,561 chinstrap chicks. By the end, when I closed my eyes, all I could see was a sea of chinstrap chicks.

After the census, Matt and I got one of a few relaxed days, which was total glory. Ever since the chicks had hatched about six weeks ago, I'd been running around, pushing away the tide of exhaustion. With a break in work, all the accumulated tiredness washed over me at once.

The break didn't last long: two days after chinstrap chick census, we did our round of chick banding. Chick banding required the entire crew at the colonies. Matt and I separated out enough bands for 10 percent of the chicks at each colony. The banding round took all six of us: two people to hold both ends of a large net near the colony and the rest to herd the chicks into the net. The net holders then walked toward each other, creating a circular pen for the chicks to stay in. The rest of us lifted chicks, banded them, then released them outside the net until we'd gone through all the bands for that colony. Securely fastened around a flipper above the right elbow, the band would almost entirely disappear into the chick down. I imagined a field tech ten years later spying one of the bands we'd just put on, marveling at the years the penguin had survived, one among the penguins that seemed now so clueless and uncoordinated.

We still had to head out to the colonies to track the few nests where parents were still dutifully tending to their chicks—unlike those chicks that had already crèched. Matt and I sat down at the skua shack in the morning, had some tea, chatted a bit, listened to a podcast. My legs felt like lead, and my neck couldn't seem to support my head. I leaned back in my chair and stared up at the chipping ceiling. Matt had moved back into main hut for a few nights because the Chilean hut was unheated, and he grumbled about not being able to sleep because of Whitney, Sam, and Doug staying up late drinking and playing a raucous game of pirate's booty, an ancient card game unearthed from the "party tote."

I hated to see him so bummed out, so I kept trying to fix things— bring a heater in to sleep better in Little Chile! Wear earplugs when you're in main hut! I always wanted to make things better, and he stared into the distance, intoning a worn "Yeah."

Though he seemed unhappy, Matt was still the friend I knew. We updated each other on the novels we were reading. We talked about our families and childhoods. We talked a lot about relationships. I'd always felt comfortable opening up to him, but in the years we'd known each other I'd grown and learned a few lessons. In the long hours at the shack the conversations waded into deeper waters. We talked about hurt and love and every kind of fear. When we headed out to do the skua show by ourselves, huddled on exposed perches, we thought about our conversations and then came back and talked about them again.

From what I knew of him, Matt seemed happiest when in a committed relationship, and he'd been through much more than I had when it came to love. While I have had a few relationships over the years, they never lasted longer than a handful of months. I was skeptical of depending on another person to be happy. I was used to marching off on my own, off to do what I wanted without negotiating or compromising. I moved fast and had my shit together; to this day, few things irritate me more than having to wait for someone else to be ready. Matt says that I have five minutes of patience a month, and once that's used up, it takes a month to build up another five.

Field affairs were lovely bouts of intimacy, but they came and went with the season. I loved spending time with Renato and missed him when he left, but I knew he would return to his life in Chile and I would move on to the next field job. Long, close, and lasting friendships had always been the most important relationships in my life, anchors in my ever-changing world. From what I saw, romantic love always seemed to require giving something up.

In the turbulent eighties in Latin America, my mother was thriving as a journalist, writing about current events, when she met my dad in the heat of Panama City. His reporting job at a news agency was more stable, so her work was the one put aside when my sibling and I came along. Even though she's brilliant and supremely capable, she found it hard to keep building her career with two kids and constant international moves. She picked up part-time gigs wherever we moved and has been a researcher, a translator, and an editor; she's worked for US agencies abroad, even as a reporter sometimes. I know she loves us. I know she doesn't regret anything. But I also know that a part of her mourns what her career could have been. She raised me to be fiercely independent. My childhood was bursting with books about strong female leads, such as Philip Pullman's character Lyra in the His Dark Materials trilogy. I loved those books.

Sometimes I feared that I'd cling so stubbornly to my independence that when someone came along and offered me something bigger, greater, deeper, I would only be able to see it as a tether. As the woolly chicks scampered around outside the skua shack, I tried to explain this to Matt. He told me it's not like that. When you love someone, and the love is full, and good, it's the best thing in your life. It doesn't feel like a sacrifice because the things you give up stop mattering as much as the happiness that comes instead. He told me that there is freedom too in vulnerability and surrender, when someone knows the entirety of who you are and loves you for it. When the person helps you become yourself.

I guess that doesn't sound so bad, I thought.

Watching my friends disappear into their relationships, I think I resented romantic love for sweeping them away from me, for demoting me, for showing me that I would always come second to this greater and deeper connection. I watched from the shore as they sailed off into love, to a place where I could not follow. I could only guess what it was like out there. When they washed up onshore after a storm, half-drowned and gasping for air, I would be there to carry them back to land.

I was also scared. Falling in love for me had only ever hurt.

We looked out at the penguins through double-paned windows, rain hitting the glass. I sipped stale, lukewarm tea and hoped that when I was wrinkled and gray I wouldn't be alone.

———

On February 9, the last few chinstrap nests I'd still been following crèched, which meant the end of daily penguin nest checks. In the morning it was sunny, but the wind was blowing hard, intermittently gusting up to fifty miles per hour and hovering between ten and twenty in between, so I couldn't brace myself well because it kept changing. Staggering, I finally got to the north of the peninsula and walked over the hill to the colonies. The second I stepped over the ridge I could feel it before I'd really seen it: overnight, all the adult chinstraps had disappeared. Chicks had been crèching here and there, a trickle, and suddenly they were all on the rocks alone. Masses of waddling chicks in little piles in the colonies. So empty. So bare.

The first day we didn't have to go to the colonies, Matt informed the crew that we'd be in the Chilean hut entering data. We snuck into a room and curled up under a sleeping bag on a bunk and watched movies all day. Dry and warm. We'd decided to do this months ago: when daily nest checks were over, we promised, we'd have a secret day off. I felt okay about it because we would have plenty of time to enter data, and no other pressing thing was on our list. It was the most rest I'd gotten in a long, long time.

All those in camp seemed quieter, subdued, withdrawing into themselves to save energy. Matt still seemed to chafe at the lack of personal space. Sam and Whitney also seemed weary, and camp lost some of the energy it had in busier times. While Matt's and my fieldwork tapered off, Whitney's and Sam's field duties piled up. Besides daily checks of their surviving puppies (twelve out of thirty), they were collecting fur seal scat and conducting "systematic surveys" around the whole Cape, which meant covering the whole peninsula on a methodical search for seals that might have been tagged as puppies. They also spent a lot of time in the fur seal lab, processing the seal scat they collected (measuring krill carapaces and saving any otoliths or squid beaks).

After our secret rest day, Matt and I switched gears and started chipping away at our data entry: putting all the data we'd collected over the season in our field notebooks into Excel spreadsheets. We did all our data entry at the end of the season, instead of throughout the season as the seal folks did. That was fine with me—I could use some time indoors. In some ways it was a relief to not have to slog out there every day and to stay in camp instead, clean, dry, and warm in camp clothes, sitting comfortably at a table working on a computer like a normal person. It was also quite satisfying to see all the data I'd gathered all season result in numbers and ratios that had a bigger meaning and context: egg lay dates, birds banded, egg weights, nest census, egg losses, hatch dates, chick losses, all the device deployments during chick-rearing, diet samples, radio tag deployments, twenty-one-day chick weighs, crèche dates, chicks banded, and finally chick census.

We didn't even go to the skua shack every day as we had since we arrived. Being done with daily nest checks was a mixed blessing. I'd been checking on the same penguins for almost four months. I wasn't used to not knowing what the penguins were doing. I felt like they were hanging out without me. It was so quiet in Little Chile, just me and a silent Matt in there all day, no ecstatic calls, no slapping flippers, no chicks mewling away.

On still nights with no wind, as I fell asleep, I could hear the ocean lapping onshore like a lazy heartbeat. Over and over and over. The ocean has licked this shore for as long as people have been alive. For as long as the moon's gravity has danced with ocean tides. I listened to it like a primal lullaby and it stayed with me as I drifted off into unconsciousness and greeted me when I woke. In, out, in, out, the water endlessly compelled to movement.

Long after our ship had left these shores, the ice would return, growing and crusting over dark waters, offering its life-giving properties to the aquatic species that made the Southern Ocean their home. Deep in the winter, in June and July, the ice would form from the coast outward, spreading over the sea until reaching its farthest extent in September, doubling the continent's size. By late October, a southern spring, the ice would begin to melt and break up, releasing nutrients to a flush of planktonic life. Ice floes would drift into open waters like the ones we saw floating near the Cape. Penguins, seals, and whales would weave through the water, gobbling up the small crustaceans that lived in swarms in the open ocean. Every year in southern latitudes the ice forms and melts, forms and melts, like the earth's lungs, breathing in and out, emitting pulses of life into marine waters.

All points recur in the rotation of our earth and we spin endlessly through them. Penguins will arrive back onshore after a dark winter, find a mate, and pick up pebbles to build nests as the snow melts in the colonies. Eggs will be laid, chicks will hatch, grow, and crèche. Fur seal puppies will be born on the beaches, wander along the hills, and wait for their mothers to return.

We are not exempt from the earth's cyclical patterns. We sleep and wake with the sun. Every day we complete the little rituals of humanity. Brushing our teeth. Eating breakfast. Changing clothes. Every day returns to the same point in the cycle. Months and years have their markers. We lap up on the shore of time, rising and falling like the tide, shifting like the sand.

Rather than a circle, cycles move in a spiral—we return to echoes of the same points, but offset, irreversibly changed by the events of the last rotation. Every day we wake up and have the ability to choose the trajectory on which the spiral moves. A wave lapping onshore seems like a small thing, but over time, the persistent motion of the water pulverizes rocks and carves away whole cliffs.

FLEDGE

13.

Mid-February

An antarctic autumn descended. Average temperatures stayed around 2°C but dipped below zero more often. The wind continued to blow dry and crisp across the Cape. In January, we'd enjoyed light from 3:00 a.m. to 11:00 p.m., and a short, dim night that never got quite dark. By mid-February, daylight went from 5:30 a.m. to 9:00 p.m. Night grew deeper and longer by seven minutes a day. Once I got up in the middle of the night to pee, stumbled outside, wrapped in my musty flannel blanket, and for the first time in four months saw a handful of stars through a gap in the clouds. The sky peeked through, jewels aglimmer. I was stunned. I had almost forgotten about stars.

With a month left to go, we were running low on milk and flour. We were running low on olive oil. The leeks, the only remaining green vegetable, were liquefying. The freshies room, where our produce was stored, was stripped bare. Some of the eggs were developing mold on the outside. They seemed intact after I cracked them open, but I threw them away regardless—better not risk it. Even the plastic of cheese containers was molding. One morning at breakfast we discovered that the carton of orange juice we'd been drinking had expired in 2015.

I stubbornly tried to summon flavor from all things stale and musty, dumping piles of old spices into bubbling pots. Our kitchen was outfitted with all the basics—four stoves, a functional oven (replaced during resupply), pots and pans, a particularly favored spatula, and a blender we had to start the generator in order to run—our solar panel chargers worked fine for our electronics but didn't have enough juice for kitchen appliances. Our remaining stores were beets, carrots, potatoes, and the ever-present cabbage. Sam, whose parents were also field-workers, had inherited a few impressive recipes that only required the dregs of a food supply. A particular winner was a salad made with cabbage, ramen, toasted almonds, and sesame oil, which we named Antarctica salad and I immortalized in my recipe book.

Whitney was notorious for inventing dishes on the fly, throwing random things together and coming up with the most wonderful combinations and diverse spreads. She was a low-key culinary genius, and I made mental, and sometimes physical, notes of her combinations. Shrimp-onion-avocado-dill, cut really small and placed back in a halved avocado skin. Chickpeas fried with sun-dried tomatoes and tahini. Desserts that came in sweet, crunchy layers.

Matt, on the other hand, was not one to improvise. He'd brought a lot of recipes with him, most provided by his three sisters, and followed them exactly, often recruiting me to chop things and fry things if he was getting stressed-out. He liked having leftovers around, so he doubled or tripled the recipes and cooked in the giant cauldrons we usually used to melt snow, filling them to the brim with some salty, soupy, calorie-delivering deliciousness. The leftovers fueled many lunches and usually lasted until it was his turn to cook again.

Deep into February, our fresh food was almost gone, and we started leaning more on cabbage and potatoes. Whoever was cooking riffled through old cookbooks with increased exasperation. Is there a recipe with just cream, cabbage, and pasta? Anything with artichoke hearts and salted almonds? Cheese and kipper snacks? Sam made the recipe on the back of the sweetened condensed milk can, except that he had to use three different cans to piece the recipe together because the cans were too rusted to be legible on their own.

I'd never cooked much with cabbage, and as I started using it more, I became impressed with its versatility—grilled cabbage wedges, leaves boiled and wrapped around a spicy filling, cut in strips and stir-fried, shredded for a slaw. But even the tricks that can be done with cabbage are limited, and after a few weeks of a cabbage-heavy diet I could barely stomach another bite. Potatoes, thankfully, never got old. Every meal with potatoes was a delight, which never went unremarked on by the crew: hums of satisfaction from Whitney, double raised eyebrows and a satiated sigh from Sam, and a silent reception from Matt, but a clean plate at the end of the meal. Perhaps something about the hardy substance of the tuber always hits the spot in barren, windswept lands. Soon, even our potato levels became worrisome.

I often admonished myself for feeling so annoyed with cabbage, given that our rations were luxurious compared with those the first explorers in Antarctica lived on. I'd found an old, faded copy of Alfred Lansing's *Endurance* in the book tote, which detailed the incredibly harrowing two-year ordeal of Ernest Shackleton and his men. I read it in February, and it made me grateful for our modest supply hut. At least I wasn't subsisting entirely on seal fat or penguin meat or a bar-shaped substance made of fat and dried meat called pemmican, a typical staple of long expeditions in the early twentieth century.

In 1914, two years after Ross and Amundsen reached the south pole and a year after Mawson's harrowing journey, Shackleton embarked, intending to cross Antarctica, from sea to sea, in what he called the Imperial Trans-Antarctic Expedition. The *Endurance* froze into the pack ice in the Weddell Sea and began drifting north. The crew lived on the ship until it gained water, was crushed by the ice, and sank. The wreckage of the *Endurance* was found 107 years later, in March 2022.

As the floe the crew was camped on split into ever smaller pieces, they packed into lifeboats and journeyed to Elephant Island—six days of navigating through pack ice. There was no human presence on Elephant Island, two hundred miles to the east of Livingston, one of the last of the South Shetland Islands. In what is still considered one of the greatest feats of seafaring of all time, Shackleton and a few others navigated a lifeboat eight hundred miles across the Drake Passage to South Georgia, where a major whaling station operated. After they landed on the south shore, Shackleton and two companions, Frank Worsley and Tom Crean, hiked for thirty-six straight hours across the island to reach the Stromness whaling station—the point at which they had departed for Antarctica two years earlier. The scene in the book where Shackleton, Worsley, and Crean finally limp into the whaling station, in tattered clothes, skin blackened by months of burning seal oil, hair long and matted, and meet the foreman at Stromness moved me to tears when I read it in my musty bunk in the South Shetlands.

Endurance was written with the help of logbooks, interviews with the survivors, and diaries. Over the months at each camp, Shackleton's crew of twenty-eight subsisted on seal and penguin, surviving thanks to the abundant wildlife of the Antarctic Peninsula. After stomaching so much fishy meat, antarctic explorers craved fatty, sweet foods. Conversation often drifted toward food they'd had in more populated places. In the journal of one of Shackleton's expedition members, Orde-Lees, there is a brilliantly indulgent fantasy:

> *We want to be fed with a large wooden spoon and . . . be patted on the stomach with the back of the spoon so as to get in a little more than would otherwise be the case. In short, we want to be overfed, grossly overfed, yes, very grossly overfed on nothing but porridge*

*and sugar, black currant and apple pudding and cream, cake,
milk, eggs, jam, honey and bread and butter till we burst, and we'll
shoot the man who offers us meat. We don't want to see or hear of
any more meat as long as we live.*

While I didn't have to hunt penguins or sleep under a wooden boat,
when I read that passage, I felt an echoing longing. Something about wind
and cold makes you crave endless amounts of delicious sugar. As the
weather grew wetter, rain spitting horizontally with the wind, Whitney
and I began crafting increasingly elaborate desserts. We made cheesecake for four straight days—sequentially, because they had to cool overnight, and if we had one ready from the day before, the odds of the cooling
cheesecake's making it through the night intact were significantly higher.
Letting things cool was not the crew's strong suit. I often pulled something out of the oven and threw it outside in the snow, with all intentions
of giving it the full recommended cooling time, but after a few minutes
Sam would say something like "It's *probably* cold enough now. I mean,
we're in *Antarctica,*" peering out the window at the dish and appealing to
my own lack of patience. We'd bring the pie or the cake or whatever it was
inside and tuck into the lavalike, gloopy mess like vultures.

Matt and I spent our days staring at computers and taking on projects at camp, such as organizing supplies, deep-cleaning shelves, and
replacing crumbling planks on the deck. He had moved back to Little
Chile and accustomed himself to the chill. I entered data in finger-
less gloves, my breath white puffs in the musty air. We had few data-
collection duties left, the most arduous of which was weighing fledging
chicks when they were fully feathered out and ready to go. Fledge, for
birds that fly, means the moment in which the chick has flying feathers
and leaves the nest, newly independent, to begin its life. For penguins,
a chick fledged when it had its full coat of proper swimming feathers
and leaped into the ocean to become a marine hunter. The chicks in
the colonies were just growing out their coat of waterproof swimming

feathers and would be fledging soon. I checked on the colonies on my skua rounds, which I was still conducting every fourth day, and which passed straight through penguin land. Once the skua chicks reached forty-eight days old (big enough to hold a band but not yet able to fly and get away), I weighed them, took some measurements, and banded them. At forty-eight days the skua chicks are considered fledged, and I no longer checked on them for the rest of the season.

Meanwhile, the seal crew pushed through their own drudgery. They were still collecting scat (seal poop) and processing it for krill carapaces. A carapace is the hard shell of a crustacean and can remain intact after going through a mammal's digestive system. Like our penguin-diet study, separating and measuring these carapaces can indicate the ages and sizes of available krill and give insights into krill populations in the ocean. The seal crew also found otoliths, that magical bone, a window into the seal's fishy prey. Sam and Whitney sat in the fur seal lab, which did not have a propane heater, bent over a microscope under layers of puffy jackets, using forceps to drag centimeter-long carapaces across a slide.

I had bought a packet of cigarettes in Punta Arenas to bring with me as a last-resort pick-me-up. I smoked about one a month. Deep in February, Whitney asked to have one, and when I slipped her the pack, much more passed between us than stale tobacco: we were worn thin, and while a cigarette is a temporary boost, it is a boost nonetheless. We do what we can with what we have. Although we had our moments, the mood around camp was weary.

With a month to go, we were all taking time alone. Somebody would go out to do something, such as count the phocids (true seals) on a set of beaches for the Friday phocid survey, and stay out far longer than it took to do the survey. The crew knew not to ask what took so long. It was that time in the season. Sometimes we needed to be alone.

We often had conversations about the things we missed. Jesse missed wearing real clothes. All Whitney wanted was to sit on a couch. Sam wished he smelled clean more than once every two weeks. We joked about getting up in the middle of the night to "accidentally" burn all the bibs so we couldn't go outside, or dropping the notebooks in the ocean so we wouldn't have to enter data anymore, or being incapacitated by a mysterious affliction, which meant we couldn't even help with inventory and could only lie in our bunks for the rest of the season, and sleep, sleep, sleep . . .

Matt and I still headed to the penguin colonies every couple of days to check on the chicks. By the third week of February, the chicks began losing their down and growing real feathers. Some chicks already looked like adults, albeit smaller and with potbellies. But most were stuck instead in the awkward "teenage" stage, covered half with dense, sleek penguin feathers and half fluffy down. Their down stuck out in many styles: Beethoven wigs, muttonchops, slick Mohawks, the sweater vest, the fluffy armpits, the full sweater, the Superman cape, the apron, the boa. When we began to see chicks jumping into the ocean, and the colonies thinning out, then we would know that fledge had begun.

On a trip out to the colonies in mid-February, I sat down in the gentoo colony by the skua shack, half-molted woolly chicks everywhere. A few adults were sprinkled through the colonies, delivering meals. Most chinstrap adults had stopped provisioning their chicks and were out foraging, preparing for their own molt. The nonmigrating gentoos fed their chicks a little longer than the chinstraps and molted later too. The colonies were quiet, save for the chicks.

I felt the rare warmth of the sun on my windburned face.

The discarded down from the gentoo chicks blew in the wind and stuck to every rock. The mucky ground was carpeted in wet feathers, ground into the mud by webbed feet. Encased in fluffy down and growing feathers, the chicks overheated quickly when the sun was out. Some lay splayed on the rocks, panting. I could hear their labored breathing all around me, a chorus of air going in and out. One approached me guardedly, investigating me with a snaky neck and large eyeballs, suspicious, yet lured by the sweet relief of my shadow. It stepped forward on pink dinosaur feet, hesitating. Slowly it scooted closer to me and ducked under my shadow, getting as much benefit from the shade as possible, and breathed relief. Its eyes never left me, scanning my form for familiarity. Leaning forward with an inquisitive bill, it nibbled at my pants and my fingers, peering up into my face, inches away, wondering, *What are you made of, shade giver?*

Down on the beach, a train of gentoo chicks waddled by a handful of irritable elephant seals. Wine-red and pale yellow algae lay strewn among the animals. When a gentoo got too close, an elephant seal, easily alarmed, jerked its head over with an expression of extreme offense, gummy pink mouth wide open, and the gentoo, startled, gave a little sideways half step, veering from its path just a bit.

I loved watching interactions between different species. Not predation, but the simple bumbling together of a breeding colony: elephant seals looking up when a troop of chinstraps walked by, a penguin hopping over a fur seal tail, fur seals approaching a gentoo colony and staring around at this nonseal world. I saw myself in these encounters: one more species navigating about the island and its inhabitants. Sometimes it seemed as if no hard distinction made me different, human, nothing like the separation I had been trained to see, the hard line between all things human and all things nature that Western culture loves to reinforce. To the animals, I imagined the scientific presence on the island was just part of the landscape. One island had steep cliffs, one island good foraging on the north side, one island strange elongated penguins running around with notebooks.

The mist was low on the water and the blazing evening sun shimmered over it. When gusts blew and picked up sea foam, it made a faint and momentary rainbow. Hard to believe that I got to be part of all this. Some days it felt like a dream.

Nearing the end of the season, I could feel myself detaching. With-drawing. I always did this before leaving. I did this as a kid before the next big move, and I did it as an adult nearing the close of a field sea-son. It was so automatic I sometimes didn't even realize I was doing it. I gently tugged at the roots I'd grown, pulling them loose from dark, rocky soil, gathering them up. So when I was physically separated, I'd come away whole. So that I'd already left when it would be time to go. It was easier. But it always left me feeling numb, and that numbness was already creeping into my days. This quiet, introspective time was like living inside what was already a memory. Like when a big wave is form-ing, and it pulls water from the shore, leaving an expanse of bare sand as it gathers force just beyond the beach. It lasts a moment, a blink, a few seconds, but that time is a pause, a stillness, a nothingness, before it all comes crashing down in a salty foam. In that moment I allowed myself to feel nothing.

———

In February, the whole crew was going on regular leopard seal cap-tures, an opportunity to head back into the field and break up the mo-notonous tasks of late season. As part of the monitoring program, we were deploying joint time-depth recorders and geolocators on the seals, which, like the devices we deployed on penguins, measured the depth of their dives and the locations of their foraging trips. The candidates were leopard seal females that were seen frequently hauling out on the Cape, so we could be more confident that they would return and we would be able to retrieve the instruments. They also had to be animals that hadn't been captured in the past few years—the small flipper tags were perma-nent and identified individuals over multiple seasons.

Doug checked the western beaches every morning for leopard seal candidates: a little over a mile to the northern beaches and nearly an-other mile down the coast. Most of the leopard seals hauled out in the west. If he spotted one, he got on the radio and let us know there'd be a capture. Once the call was made, everyone mobilized. Matt and I leaped out of Little Chile's wooden chairs and rushed to main camp to pull on field gear and grab all the leopard capture stuff. Sam and Whitney mobilized from the fur seal lab, stopping their own data entry. The captures were a huge production. We needed so much equipment: a massive metal tripod, a tarp for weighing with metal rods at either end, the winch, the dart gun, the sedative kit, the tape measure, the instruments. We had to get out there as soon as we could because we didn't know how long the seal would stay onshore.

Doug would be waiting for us on the leopard seal's chosen beach. As the primary medical officer, I monitored the respiration and heart rate of the animal. I crouched behind rocks and snuck a look at the enormous seal with binoculars, counting breaths by the rise and fall of the chest or the opening and closing of the nostrils, depending on the angle and which I could see better. Once I had a baseline respiration rate, I nodded at Doug, who raised his dart gun and shot the leopard seal with sedatives. The seal usually reared in surprise, woken from its nap, but soon lay down again, subdued. We crouched behind rocks and waited until the drugs kicked in. After ten minutes, Doug approached carefully and quietly, patting the animal to see how asleep it was. If it didn't respond, he inserted a needle into the animal's spinal column and administered an IV loaded with more sedatives. Unlike gas anesthesia for fur seals, which made them unconscious, the sedatives we used for leopard seals just made them really sleepy and calm.

Once the IV was in, the rest of us snuck up with all the gear. Sam and Matt took the seal's length and girth with a measuring tape. Whitney took biological samples by poking it with a small sharp tube that took a plug of skin. I read respiration rates every five minutes to make sure the seal was still breathing normally and took down all the data from the rest of the crew. Sam and Matt attached a heart rate monitor for the duration of the capture and glued the geolocator to the animal's back with epoxy.

The last thing we did was to roll the leopard seal onto a tarp to weigh it. Tucking the tarp under the enormous, bulging animal took all six of us, as we heaved to roll it just enough to tuck the tarp under it halfway, then circled around to push the animal from the other side just enough to grab the roll of tarp from under it and spread it out. The tarp would be secured to a pulley and a rotary hand winch, which Sam cranked until the seal was lifted off the ground, supported by a massive metal tripod, which was a pain in the ass to carry around the hills of the Cape. The first female we captured weighed almost eleven hundred pounds. Once the animal was weighed and we'd rolled it off the tarp, Doug administered the reversal drugs while we cleared from the beach, crouching low and moving quietly. The leopard seal would become alert within minutes and would not be happy, not happy at all. While the crew packed the gear out of sight, I crouched behind a rock to take one last respiration reading, counting breaths with my binoculars, and then we cleared off. The whole process usually took about an hour and a half.

Leopard seals are extremely hard to study because they are so solitary and mostly live on ice and in the open ocean. Until 2010, few were found on land in the summers at the Cape. Not much is known about them. Mostly females have been studied to exclude sex as a factor that might be causing differences in foraging patterns between individuals.

Doug was largely steering the leopard seal research at the program, as he'd done critical leopard seal research as part of his PhD with NOAA. This season, Doug decided to include two males in his sample. Nobody had ever put these kinds of devices on leopard seal males before. And if we were going to do a male, we had to do two because each leopard seal has individual hunting tactics and preferences, and any information we got from one device deployment could be specific to that seal and not relevant to males in general. For example, some leopard seals liked to hang around the seafloor, hunting demersal (seafloor) fish. Others waited offshore from the fur seal colonies for puppies to wander in the water. After the captures, all we could do was sit tight and hope those animals would show up again in a week so we could do the captures to retrieve the devices. We did eight captures in all, one deployment and one retrieval capture for four animals: two females and two males.

Doug was the first to capture leopard seal foraging behavior on film, using cameras attached to the animal's back. Doug's PhD was a joint project with the Scripps Institution of Oceanography, NOAA's Antarctic Ecosystem Research Division, and National Geographic's Remote Imaging, which had developed the "crittercams" he deployed in the 2013 and 2014 Cape Shirreff field seasons. Through the footage and attached diving and GPS data, Doug found that most seals have their own specialized hunting tactics. The seals tailor their hunting to capture their prey of choice—be it penguins, puppies, fish, or krill. He found that they cache food on the ocean floor (penguin and seal carcasses) and steal food from one another. The larger seal wins the carcass.

One night Doug told us about getting a camera back from a leopard seal at the Cape during his PhD research. He was so excited to watch the footage, feeling as if a window into an unknowable otherness had suddenly opened. He sat on his bunk and watched the footage on his laptop for hours. Some things surprised him, excited him scientifically, such as two leopard seals fighting over a carcass—the first evidence of intraspecies competition. Hours later, late at night, Doug was still pinned to his screen, experiencing something like what the seal experienced in the days it carried the camera. The seal was swimming in an ambiguous blue, a marine world that could have been anywhere in the waters around the Cape, then broke the surface of the water. Through the screen, Doug saw a rocky beach, fur seals, and beyond it . . . camp. Us. Self through the eyes of the other. A chill ran over him. He slowly closed his computer, thinking, That's enough for tonight.

14.

Late February

Toward the end of February, the chinstrap chicks began to fledge. In the morning they gathered at the shore with the few adult penguins that were left in the colonies to leap into the sea. The young penguins were unmistakable in the crowd—not just because they had a dark smudge around their eyes, a stubby tail, a bluer back, and were a bit smaller, but also because of their air of uncertainty. They showed up at the dawn exodus in their fresh tuxes, all sleek and crisp, looking around with wide gray eyes and no clue what the ocean was, but with an intuitive sense that all these other penguins were going where the food was and it was best to follow along. All the adults around them had the glazed-over morning look you might expect from a New York subway at 7:00 a.m., moving forward in a sleepy stream of bodies. Matt and I imagined the fledgers like clueless tourists: "Erm, excuse me, sir, I was told to arrive here at this time. Um, I'm new around here. . . . Is this the line to the Ocean?"

The beginning of fledging meant that Matt and I had to go out at peak exodus, right at dawn, and weigh fledging chinstrap chicks before they swam off. Fledge weights tell us about the chick's body condition and fat stores when it headed out to sea, and thus how much time it had to find food before it starved.

Matt and I took turns waking up at ungodly hours and getting to the colonies at first light. I'd sleep in the Chilean huts when it was my turn, so I wouldn't wake up the rest of the crew when I headed out. The Chilean hut had two bunk rooms separated from the kitchen by a luxury we did not enjoy in our own hut: doors. I slept in the room Matt was not occupying and set my alarm for the wee hours of the morning.

Getting up at 3:45 a.m. was a struggle. My sleeping bag represented everything warm, safe, dry, and comfortable, and out there was darkness, wind, cold, wet, work. All the clothes I had to pull on were clammy and frigid. But once I hauled my reluctant ass out of bed and out the door, the journey had its own rewards. The late night was deep, and new, even if it came with a blast of freezing wind to the face. It had been a long time since I'd seen night. With it came stars, and in predawn, if it was clear, a handful of them glinted down on me like old friends.

Under the light of a dim orange glow in the east and a pale half-moon, I made my way across the flats. All the water that drained from the hills froze overnight in a film over the ground and crunched under my boots. The sun would break the horizon around 6:00 a.m. When I hiked out, it had to be bright enough to differentiate a rock from a seal, but not much more than that. The navigation was by muscle memory—I had walked this route twice daily for many months.

In the colonies, penguins grouped on the shore, gathering, then leaped into the ocean. Chicks joined the throngs, heading out to sea for the first and most perilous year of their life. When I caught one to weigh it, it looked up at me with wide eyes, measuring and calculating. Their eyes darken as they age, but when they are fledging, they have beautiful light gray irises.

Once the chinstraps leap into the sea, they will be in the ocean until they return to the colonies to breed, after three years. Some may return earlier during the breeding season, like the few juvenile penguins I occasionally saw in the colonies and on the landing beaches below. Most of their time will be spent swimming in the Southern Ocean, hunting krill. They rest, preen, and live on the water—chinstrap penguins have little use for land, except to breed and to molt.

Matt and I traded off fledge weights for a week—protocol indicated that we should aim for around two hundred fledge weights. Walking all the beaches below the colonies, I managed about twenty or thirty in a morning. It was much easier with two people, two penguin-catching nets, and two people to weigh and record data. The seal crew helped us out—Sam headed out early one morning with Matt, and Whitney gave me a hand one morning.

After running around doing fledge weights in the early morning and then leopard seal captures in the afternoon, I was tired. It was much more fieldwork than I'd been anticipating for this point, and my nerves were worn and ragged. The island was always both beautiful and brutal, welcoming and hostile, still, merciless, and sublime.

Living in field camps can be a crazy world of two extremes. You're shown the farthest, most remote side of the earth, and it's huge and amazing, then you're left out there, in the huge and amazing, in the endless wind, with no pelt, no coat of feathers, no layer of blubber, you, a human, on human legs, naked, cold. All your carefully constructed layers get torn, peeled, tossed around in the wind, and through the tatters shines a faint ray of antarctic light onto the very center of you, now exposed, now lit, and it glints ever so delicately before you bind all your humanity again in a shroud around it. Humans didn't evolve to survive in this ecosystem. The island demanded transformation. You had to be a rock. You had to be the wind. You had to be a penguin.

———

On one of our last fledge weights, both Matt and I got up early and hiked out so we could deploy five GPS trackers on fledging chinnies. The PTTs activate when in contact with salt water and beam their location to a satellite that transmits to NOAA's Antarctic Ecosystem Research Division headquarters. In the few days following the deployments, our boss, Jefferson, monitored the penguin's movements from San Diego. He emailed us to let us know that a couple fledgers were rounding the curve of King George Island, fifty miles away, and another had just headed straight into open ocean and was already ninety-five miles away. Matt and I felt like nervous parents whose kids have just gone off to college: proud, sad, relieved, but mostly anxious. It was hard to imagine those little chicks we watched grow up out in the cold, unforgiving world. Did they have everything they needed? Were they healthy enough? Would they figure out how to find krill before they were hunted? We fussed and dithered.

While the chinstrap chicks had to contend with an abrupt, literal leap into adulthood, the gentoo chicks enjoyed a gentler transition. The young gentoos took short trips out at sea, learning to forage, while still getting additional meals from their parents. They ran everywhere in little gangs, splashed around in the shallows, and snorkeled in tide pools, tail bristles sticking straight out. As the days went on, they became increasingly independent. Penguin researchers, including Jefferson, have suggested that this is one of the reasons gentoo penguin populations are increasing in the Western Antarctic Peninsula, while chinstrap and Adélie populations are decreasing. The more gradual transition buffers the chicks from the difficulties of finding krill on their own in a krill-limited system. The gentoo training period becomes more significant with less krill available and makes the species more resilient to changes in the climate.

Since 1994, gentoo penguins have expanded their breeding range thirty-seven miles southward. The puzzle is in figuring out why these trends are so different from those with chinstraps and Adélies, when adults seem to survive the winter at the same rates, and the breeding success (proportion of eggs that lead to crèched chicks) is also similar across species in any given year.

The key differences could lie in their varying winter habitats. Chin-straps swim off into the open ocean and generally avoid ice. Good years for chinstrap juveniles tend to be correlated with years with an abundance of large, adult krill, which is mostly found in swarms in the open ocean. Adélies are ice dependent. Years with an abundance of juvenile krill favor Adélies, as juvenile krill live and hide under the ice from which Adélies hunt. Gentoos are nonmigratory and stay around their natal colonies over the winter, hopping onto the beach on occasion, with an extended transition time that buffers them against some of the vulnerabilities of their first year. Gentoo penguins also have a much more flexible diet, eating fish as well as krill.

In 2020, Jefferson published a paper with the results of the tracking we did in 2017 and 2018, showing that high juvenile mortality was likely the bottleneck to chinstrap and Adélie population growth. Juvenile penguins, having just entered the ocean for the first time, are vulnerable, inexperienced, and disproportionately impacted by an increasingly krill-limited system. Chinstrap parents simply stop feeding the chicks and leave them to their own devices. Initially, juvenile hunting efficiency is low, and they have less energy to avoid predators. Within sixteen days, 73 percent of the tags we deployed on juvenile penguins at the Cape along with tags put on King George Island (Adélies, chinstraps, and gentoos) had been lost. Most of the juvenile penguin deaths occur within three weeks of leaving their natal colonies. We know that they were probably chomped by leopard seals because the tags disappeared before the penguins were likely to have starved.

The reduced availability of krill could be a result of both climate change (less sea ice) and increasing competition by the krill-fishing industry. The Association of Responsible Krill harvesting companies (ARK) is a coalition of 85 percent of the krill-fishing industry that operates in the Southern Ocean, joined together to better coordinate sustainability efforts. ARK has made voluntary commitments to stay twenty-five miles away from summer breeding colonies. Voluntary restrictions can be agreed upon much more quickly than passing a provision unanimously through CCAMLR. The krill fishery is shifting toward being more active in the autumn and winter months, when krill contain a higher concentration of valuable oils and less algae, and when the seal and penguin breeding season is over. This is significant because juvenile penguins are competing directly with the fisheries for krill right after they fledge and are just learning to find food.

CCAMLR's approach to managing fisheries is "precautionary": with uncertain data (it is difficult to estimate how many tons of krill exist in the Southern Ocean), restrictions err on the side of caution. Overall catch limits are less than 1 percent of the estimated stock of antarctic krill. CCAMLR has also set up subareas with their own limits to avoid targeted krill extraction in regions that are ecologically important for krill predators. Nevertheless, there is significant overlap between where the bulk of the krill fishery operates and where predators, such as penguins and whales, forage. Jefferson, the seabird research lead, published a recent paper with other collaborators about how CCAMLR's management areas operate at a scale that is much bigger and coarser than the scale at which predator-prey interactions occur. The paper concluded that even within CCAMLR's subareas, high fishing intensity just offshore from important penguin breeding colonies, such as Cape Shirreff, could still have a significant impact on these predators despite the overall precautionary approach of the fishing limits. The maximum catch allowed for the Antarctic Peninsula, a crucial habitat for a lot of breeding populations of penguins and seals, is often reached by the fishery (but not exceeded).

In the South Shetlands, Adélie and chinstrap penguins, both krill dependent, have declined by more than 50 percent in the last thirty years. Chinstraps are an ice-avoidant species, while Adélies are an ice-dependent species. Decreasing numbers of both species indicate that population declines are not linked to ice availability, which would impact only Adélies, but rather a decrease in the availability of their main food source: antarctic krill.

While it's indispensable for penguins, krill has been a hard sell as a part of our human diet. Back when Russia was the USSR, introducing krill there as food—krill paste, peeled krill tails, and canned krill—had little success. The krill fishery was established by the USSR simply because of the massive abundance of krill, not necessarily because people wanted to eat it. Today, most krill are turned into fish meal, which is used in China as a high-value additive in aquaculture feed or as fertilizer in agriculture. Ten percent of harvested krill are processed into krill oil, a figure that is likely to increase. Krill oil is sold as an omega-3 supplement and marketed at a higher premium than generic fish oils because it contains astaxanthin, the compound responsible for krill's pink color. Astaxanthin is peddled as "nature's most powerful antioxidant" and has been suggested to have health benefits, including as an oral sunscreen, although the field is still in its infancy and most of the research has been inconclusive.

In the middle of February, I saw two huge factory trawlers in the far distance sucking up giant swarms of krill that could otherwise have been hunted by a chinstrap raising two chicks or a fur seal mother or survived to breed more krill. The trawlers lingered on the horizon, stacks pumping out smoke, greed-driven industrial forces, products of late-stage capitalism, perverse metal monstrosities in one of the world's most enigmatic ecosystems. I stood in a melting patch of snow, a human, on an island with some penguins, armed with my data book and a pair of binoculars. I gave the ships a middle finger, sighed, and turned to go inside.

I couldn't denounce the fishery entirely because my job was tied to its very existence. CCAMLR, the international administrative body, was set up to run studies mostly to inform the regulation of the krill fishery, although from the beginning Mike and his seabird counterpart knew that climate change would be a central part of their studies at the Cape. NOAA was the national US agency that ran the research programs that aligned with CCAMLR protocols, and whose data would then be taken to the annual CCAMLR meeting in Hobart, Australia. While most fisheries are managed based on population analyses of a single species, CCAMLR was unique in that it was the first international fisheries agreement to incorporate an ecosystem approach. The ecosystem approach means it wasn't just about "how much was out there," it was also about the other animals that depended on krill, its place in the food chain, climate change, and the other factors that might be impacting populations.

I often wondered, staring out at the Southern Ocean's turbulent surface, what the science would look like if it weren't tied to regulatory needs. Not all the research that was carried out in camp was directly related to CCAMLR's data-gathering requirements—but CCAMLR protocols were the backbone of the ecological-monitoring program. Without the krill fishery, would we find other reasons to set up monitoring stations and field camps? What would we work to discover and understand? How would we relate to these fragile and remote ecosystems? Now that climate change is poised to threaten the very foundation of these vibrant shores, what is the value we seek to save? Is it economic? Scientific? Or something more?

Out in the ecosystem itself, autumn tightened its grip and it began to feel much like winter. In late February we got a few intense, crazy storms. In the morning we woke up and the whole Cape was covered in snow. Snow was crusted over all the doors, the walks, and the deck, with the temperature dipping to –1°C, wind howling with gusts to forty-two miles per hour.

The snow blew horizontally, following the wind. Sun lit it ablaze. Wind sculpted form from flatness, macro-drifts and micro-peaks, ridges, and pileups behind pebbles so it looked as if every rock had a bright white shadow. Overnight everything froze in place. The penguins painted a delicate line of blue footprints on stretches of white. Cloud shadows made dark shapes on the ground. Matt and I holed up in the Chilean hut and typed away, while the snow piled up against the windows and wind tried its hardest to tear off the roof. We were finally done with fledge weights. We burrowed deeper into our data hovel except for when Doug spotted leopard seals with the devices on the north coast and we all trooped out for a retrieval capture.

Data entry could be tedious, but seeing a season's work added to the long-running data set was gratifying. Each nest was entered into a line in a spreadsheet—after daily visits for four months, we got one line: lay date, number of eggs, hatch date, number of chicks, crèche date, fledge weights, and any deaths. Scrolling up, I got to see the data from past seasons, all the way back to the late nineties, hundreds of lines representing thousands of hours of observation, all summarized neatly in these squares, to be analyzed along with all the other spreadsheets all over Antarctica filled in by similarly grubby fingers in damp field camps. A small army of field techs, I liked to think. We were isolated from one another by the nature of our work but unified by the data, the lines of graphs like tethers connecting our dots to the dots of others, painting a picture of the ecosystem to which we tied our lives.

I thought about the skills it took to turn this data into papers, to draw conclusions from the sea of numbers, to learn about the ecosystem through graphs and P values. I was pretty sure that it wasn't my calling—but entering the data, I thought of what would come afterward, when the papers were published and the diplomats gathered to make policy decisions. I wondered how this learning drawn from ice and sea worked its way into our laws and policies, into our culture, into art. I wondered if I could find a home in the space after the research has been published—to help learn from it, help interpret it, help apply it.

During the day the Chilean hut was Matt's and my office, but at night it became his lair, his space away from the oppressive presence of other people. The crew still gathered every night for dinner. Sam had been looking for his earphones for a month. Whitney never stopped moving.

Often the last to sit down for dinner and the first to get up, she always had something to busy her hands with—cleaning, organizing, repairing gear. Our wine and margarita consumption had accelerated, and the edges of the season were approaching enough to reinfuse camp with some energy and joviality.

Doug, a great entertainer, did wonders for camp morale. When he launched into one of his stories while we puttered around before dinner, usually about some shenanigans he'd got up to with some "lovely lady friend" or other from the past, we'd all settle where we were, listening attentively. Everyone moved more quietly, stifled sounds so as not to interrupt. Sometimes he'd pause in the middle of a story and absentmindedly rub the growing scruff on his chin, peering at the ceiling, lost in the memory, and chuckle to himself—silently, all breath and shoulders—having forgotten us entirely. He has an expressive face, marked by huge eyes, and he delivered punch lines with perfect timing. I was sure he'd told these stories dozens of times to generation after generation of field crews, but rather than their feeling automatic, they were polished from the repetition, details carefully chosen, characters vividly described.

One of the many times Matt skipped dinner in main hut, I walked down to Little Chile with a plate of food, complete with our latest sweet, crumbly creation, and entered his den with caution, leaving the plate on the wooden countertop. "Matt?" I called into the darkness, the faint blue light of his phone the only pinprick in the dark, musty room. He was doing a crossword puzzle. "Matt, I brought you food, you should eat it," I called into the pit of solitude.

"Thanks," I heard back, a toneless, mechanical response lacking all human affect.

I could tell that this season had been hard on him. After more than a decade of remote fieldwork, he seemed on his way out. Faced with what I felt was Matt's slump, I yo-yoed between compensating for it, trying to radiate positivity, and putting aside my energy to just sit with him. On the last season we'd worked together, on St. George, he'd had his own room in our bunkhouse to hole away in. The islands in Maine where he'd spent many summers and falls had a hut for cooking and eating, but everyone had separate personal tents to sleep in. In other field seasons, there had been a little space that was separate, and his— but not in this one.

I couldn't help but wonder if I kept doing fieldwork simply because I liked it, would I be over it someday? After ten years of fieldwork, would I wish I had invested more in my life outside the field? Was I projecting my own fears onto Matt's simple need for space? Going through the rush of feelings that all endings bring on without Matt's happy companionship felt as if I'd lost some of my footing—as I drew closer to the barren hills, the distance between us widened.

One afternoon as we typed away in Little Chile, Whitney came down to visit us, beer in hand. She told us that a device they'd just recovered from a fur seal had been faulty and hadn't recorded any data. This happens sometimes; wildlife work can be unpredictable. But it also means that the whole ordeal they put that female fur seal through—a disorienting capture, gas anesthesia, a device glued on her back, then another capture to recover the device—was for nothing. The stress we caused our study animals never felt insignificant, not to us and definitely not to them. Each disturbance was calculated to be as minimal as possible. Captures caused a lot of stress and disruption to our study animals, unlike our walking through a colony or standing near a harem, but they were the only way we had of getting data on foraging patterns. These were the tools we had to protect the ecosystem we worked in: a tiny box of circuits made a world away and stuck onto an animal's back. The stakes: what felt like the future of the Southern Ocean ecosystem.

The tireless work of our predecessors helped support the creation of two Marine Protected Areas (MPAs) in the Southern Ocean where commercial fishing is banned. Back in 2009, CCAMLR member countries agreed to establish these with distinct boundaries based on the best available science. The Ross Sea MPA, created in 2017, is the biggest

MPA in the world: 1.55 million square miles, or almost six times the size of the state of Texas.

Three additional MPAs have been proposed: one covering the Western Antarctic Peninsula, including Livingston Island, put forth by Argentina and Chile; one covering three areas off the East Antarctica coast, put forth by Australia, France, and the European Union; and one in the Weddell Sea, east of the peninsula, put forth by Germany and the European Union.

The evidence supporting the establishment of these proposed reserves often includes data from geolocators. A collaborative study published in 2020 combined tracking data from 4,060 individuals from seventeen different bird and mammal species across the continent, including penguins and seals from the Cape, to determine Areas of Ecological Significance (AES) in the Southern Ocean. The paper has a massive cohort of eighty-two authors spanning ecosystem research across Antarctica. They found that the areas preferred by multiple predator species were also areas with the highest fishing effort, as measured by hours fishing. One key AES almost entirely encompasses the Scotia Sea, which is just east of the passage between South America and the Antarctic Peninsula and veers over to cover the South Shetlands. Current MPAs in the Ross Sea and the South Orkney Islands, established in 2017 and 2009, respectively, already cover 27 percent of the AES identified in this study. Establishing the three MPAs that have been proposed in CCAMLR would bring this figure to 39 percent.

For eight straight years the proposed MPAs have failed to muster the unanimous vote required by CCAMLR to pass the measures, a difficult task with twenty-six member states. CCAMLR's headquarters are in Hobart, Australia, where the annual meetings are conducted in the convention's four official languages: English, Spanish, Russian, and French. Working groups and subcommittees meet throughout the year to tackle specific issues. The big decisions must be agreed upon by all member states in the annual gatherings. China and Russia, two of the biggest players in the Southern Ocean's fishing industry, have yet to agree to the proposed MPAs. The evidence in support of these MPAs is clear. The nature of science, however, will always allow room for doubt—we could always be more sure, we could always have more data, openings that are seized upon to divert and delay decisions that might simply lack political will.

So Sam, Whitney, Matt, and I run around on an island rock, collecting more data.

Whitney's face was anguished as she told us about her frustration and disappointment. How can we allow for some devices not to work? They all need to work. This all needs to serve a purpose. Otherwise, what are we doing? We talked about the weight of disturbing the wild species we'd grown to know so well. Instead of trying to make her feel better, Matt nodded as she spoke, acknowledging her frustration and anger. That was all she needed. And maybe that was all Matt had needed from me in these long weeks. He didn't try to fix it because sometimes things can't be fixed, and that's okay. And there was Whitney, the veneer of optimism dissolved. It was the first time I ever saw her truly raw and downcast. I felt closer to the two of them than in any of our joyful moments. Joy was easy. Four months deep and a faulty TDR was not.

15.

Early March

"End of science" was a late-season milestone when all our data collection was completed and we dedicated ourselves full-time to closing camp. Matt and I were almost done with data entry, but the seal crew was still in the throes of a series of extensive Cape-wide tagged-seal surveys and would need more time. The last data-collection duty for the seabird team was a survey of penguin carcasses on the beaches. When the chinstrap chicks fledge, many are caught by leopard seals who are waiting just offshore for the occasion. The leopards turn the penguins inside out and eat the fleshy bits, leaving the skin, feathers, and skeleton to float off and wash onshore. Despite the shifting currents at the Cape, counting the carcasses that wash up can give an indication of levels of predation for that season. In the last ten years, the count had been somewhere between forty to eighty total carcasses. Since we banded 10 percent of the chicks, usually four to eight bands are recovered during this effort.

On an overcast Thursday, we planned to walk the edges of the beach to count the carcasses. I hiked out to Media Luna beach, while Matt hiked out to the penguin colonies to the north so we could start in opposite directions and meet in the middle.

I picked my way along the rocks and stopped at every carcass. The beaches were littered at rates I hadn't expected. Most were chinnie chicks, with a few chinstrap and macaroni penguin (who nest in nearby islands and occasionally visit Cape Shirreff) adult carcasses as well, and almost all leopard kills. Most had the heads ripped off, most were inside out. I bent down to turn them outside out again to check for bands or geolocators. There were so many dead chicks tangled up in the algae, foamy surf lapping on their soaked feathers. When I turned a corner toward a little beach pocket framed by ridges, I found an actual pile. I picked through the pile, carcass by carcass. My thick rubber gloves smelled like flesh.

The beaches looked like the aftermath of a war. Carcasses were stuck in piles of seaweed, wedged under boulders, half-buried in the sand. Most were chinnie chicks. My heart sank a little further for every carcass I pulled out. Matt called me on the radio after the first two hours to check that I was seeing the same staggering volume he was. We spent hours and hours out there, counting the dead.

By the time we met in the middle, our count was 741 (we'd counted 3,562 chinstrap chicks during census). A total of 741 chicks from nests whose parents laid the eggs of those chicks many months ago from energy reserves saved up in the frigid winter, whose parents worked hard traveling from sea to shore, hunting krill in the turbulent Southern Ocean to bring their chicks a meal every day, two adult penguins trading nest duty, for months, the chicks' real feathers growing in, until finally they jumped into the ocean to begin their new aquatic existence—just to be eaten by leopard seals shortly afterward. Four of the five geolocators we'd attached to the fledging chinstraps had already stopped transmitting, which means the birds died and the transmitters sunk into the deep blue.

Matt and I sat in the skua shack afterward, cleaning chinstrap feathers we'd collected for isotope analysis, the sea of carcasses we'd just waded through weighing down the air between us. Besides massive numbers, the most surprising part of carcass count was that we found no bands. Zero. We'd banded 10 percent of the young fledgers. Even if only twenty of those carcasses were from the colonies we had been monitoring at Cape Shirreff, we would have expected to see at least one band.

If they weren't our chicks, where did they all come from and how did they get here?

We sent a frantic email to our penguin research lead in San Diego and waited two, three days for a response, feeling alarm, interest, and grief. After a few emails back and forth with him, Mike, who was still in San Diego, speculated that the washed-up carcasses probably came from one of the other colonies near Livingston Island, washed up from a confluence of currents that happened to direct the dead our way. A number of smaller islands around the South Shetlands have huge chinstrap penguin colonies—the Zed Islands, twenty miles east of Cape Shirreff, host a chinstrap colony of twenty thousand pairs; Stoker Island, twenty-nine miles away, has just below fifteen thousand pairs—big numbers compared with the three thousand breeding pairs we'd counted at the Cape that year. Possibly those colonies, many thousand strong, blanketing distant cliffs, were able to withstand the deaths of the penguins that washed up on our island. Still, the image of beaches littered with carcasses was a stark illustration of the predation these animals faced in the first year of their life.

I held out hope for the ones still out there. Fledging chicks and other penguins heading out to sea use a tactic called predator satiation, in which they all jump into the water at the same time, flooding predators, who can only hunt a few individuals as the crowd passes. A few must have made it past. The leopard seals were like gatekeepers to their future, the first big test of penguin-hood.

The last leopard seal with a device hauled out near the penguin beaches, and we hiked out to meet him. Carrying the unwieldy gear, we walked in single file, like ants threading their way through a rumpled sheet, hills rising from rocky earth on either side of us. During the capture we plucked precious troves of data from the animal's broad back. Before Doug gave the leopard seal the reversal drugs, Sam, Whitney, Matt, and I all took off our gloves to touch him, feeling his warm velvety fur and running our fingers along the scars on its side, the fleshy folds of his armpit, the hard ridges of his flippers. How many people have touched a leopard seal? He smelled musky and salty. Our soft sea monster.

———

The gentoo chicks, fully feathered, moved closer to the shore and spent time dipping and playing among the tide pools and going on short trips to sea. Some gentoo chicks still received meals from their parents, a luxury that would soon taper off as the chicks gained skill and confidence in hunting their own food. The chinstrap chicks were gone from the colonies, having leaped into the ocean and fledged. The chinstrap adults were out at sea, preparing for their own molt.

Most flying birds molt one feather at a time—on one side and then the other—to make sure that they can still fly while replacing their feathers. But penguins undergo what is called a catastrophic molt, in which they replace all their feathers at once. If it sounds dramatic, that's because it is. To prepare, they spend days at sea hunting and bulking up in anticipation of their molting fast on land. The penguins won't be able to forage until their new coat of feathers is seaworthy. I could always tell which penguins hopped out of the ocean to start their molt because they looked huge. Tiny heads on beefy bodies, padded and ready to go.

The molt takes about two weeks, and during that time the penguins, hunched and scowling in apparent discomfort, just stand there as parts of them blow away. Breeding pairs molt in their nesting spots, and non-breeders molt on the rocks near the beach.

The first thing to fall out is their tail bristles. Then the new feathers begin to grow in, pushing out the old feathers, and before the old feathers have fallen out, the penguins look inflated, like blown-up puffer fish. The absence of a tail makes them especially round. Their old, faded feathers come off in clumps, coating the ground around them. All the sharp edges of their tuxedo get blurry and bumpy, the black streaked with white.

The penguins expend so much energy fasting and molting that by the time their fresh coat of feathers is ready they look as skinny and shrunken as fledging chicks. I could see the ridges of their ribs under their shiny new coat as they waddled back to sea, hungry for krill.

Walking by the molting colonies, I thought about how the first time I'd come upon this sea of penguins they had all looked the same to me. Over the months of daily observation, I'd learned to distinguish between them: the fledging chicks, the skinny ones versus the healthy ones, the adults returning to molt, the adults molting, the adults that were done and headed back to the ocean, the nonbreeding juveniles that had returned to the colonies just to hang out, the females, the males, the breeding pairs, the ones that always slapped me when I passed by. By the end of the season, the sea of penguins was a nuanced population of individuals and personalities. I felt that I had learned to see.

So many penguins were molting that it rained short white penguin feathers. Like water, the feathers gathered in low places, forming patchy streams and pools of white. Like snow, they followed the whims of the wind, whirling through the air and piling up against rocks. The colonies were dusted with the discarded bits of a penguin year.

Nearing the end of the season, camp underwent its own molt as we began clearing away the clutter we'd surrounded ourselves with. Jesse tore into main hut with zeal and none of the sentimentality of an old-timer, it being only her second season at the Cape. She'd open totes tucked in dark corners and grow increasingly frustrated at the quantity of junk that had been stashed in the hut over so many decades. She unearthed cassettes loaded with instructions for outdated machines, fumed at the pointless bits of plastic in the party tote, filled trash bags with a sea of little moldy objects.

Doug was infected by her purging efforts. One night while we puttered around before dinner, I asked if anyone had seen the rubber-band ball, because I hadn't in a while. Doug casually said, "Oh, I threw it away." Sam and I stared at him, shocked into speechlessness. Doug, who hadn't been here when it was made—rather, brought to life—who hadn't been here when Sam and I had bonded with it, who didn't know it was valued, cherished even, looked at us with an increasingly bewildered expression. Brows furrowed, confused, he was quick to justify himself: "The rubber bands were really brittle. They kept breaking and the ball kept rolling around everywhere and shedding useless rubber bands all over the table." Sam and I replied, "Oh, okay, yeah, I guess so," and were crestfallen. Our friend, so easily discarded. Doug apologized, bemused but sensing that it was the right thing to do.

After that, Doug made sure to ask us before he threw away anything that seemed like trash but could have secret sentimental meaning. Is it cool with you guys if I throw away these paper clips? Is there a reason a penny has been on this shelf for two months? Does anyone need this note card with an abstract doodle on it? Is it valuable? Does it mean something?

Out in the colonies, Matt and I cleaned, inventoried, and winterized the skua shack. We waited for a sunny day to wash the floor to increase the chances of its actually drying. We splashed bleach, cleaning solution, and rainwater on the floor, scrubbed, and squeegeed thick, muddy pools out the door. After we washed the floor twice and rinsed it once, the water that came away was still a murky brown. It worked marginally: for the first time we could see the grain of the plywood and marveled that it had been there the whole time.

Over the next few days, we recaulked the seams in the skua shack plywood, cleaned the water barrels, put away the rain gutters, and splashed all surfaces with a good dose of bleach. We hauled empty pro-pane tanks back to camp on frame packs and hauled out full ones for next season. We carried back our penguin clothes sealed in big trash bags and threw them in the trash pile in camp. I kept having to recaulk a seam on the outside of the shack near the floor because the curious gentoo chicks would grab an end and pull the whole string of caulk out before it dried.

Matt and I indulged in our own ending ritual aside from all the closing-camp duties. We had been floating the idea of doing tattoos all season. I'd brought my little kit of ink and needles to the Cape, stashed with my socks above my bunk. I mentioned to the seal crew that we were going to do tattoos, in case they wanted one. Whitney decided to have a couple of antarctic circumpolar-current arrows added to a tat-too she already had. They both came by the shack a few days before we boarded it up for the winter to a tattoo appointment. I wondered if I was the only tattoo artist in Antarctica.

When the seal crew ambled in, Matt smiled dutifully and made his exit. Four people in the hut still seemed like too much for him. Sam, Whitney, and I poured some wine because fuck it. The propane heater was on, and Whitney hopped on the table while I sterilized everything and opened a new needle. I wiped her skin with an alcohol pad and started poking. Sam was splayed on a chair nearby, supervising.

I asked Whitney about her family to distract her from the sting. She was pale, and in the middle of our conversation she said, "I think I'm going to pass out." I paused, looking at her with alarm. She suddenly pitched forward, and before I could react, Sam was there, catching her. He picked her up as if she weighed nothing and carried her outside for some air. I will always remember that image, of Whitney, pale midriff exposed, fainted, and Sam, carrying her like a doll out the skua shack door. The force of her personality was such that it was easy to forget that she was physically small. Unconscious she was like a bird.

Recovering quickly after a few seconds, she returned to us, taking great gulps of air while Sam and I hovered (as the medical officer my hovering was particularly anxious). Sam and I figured they had been dehydrated when they got to the shack, and the warmth, wine, sitting on a table, head up high where all the heater fumes were, combined with the sharp stabbing pain of a tattoo, must have done it. I turned off the propane, opened all the windows, refilled Whitney's water bottle, and finished the circumpolar current on her ribs. "Well," she said, beaming, clipping her bib back into place when we were done, "that was exciting." She was her old self by the time she left the shack, and Sam and I looked at each other relieved because, holy shit, she scared us there for a second.

Undeterred, I tattooed an Aztec symbol of wind on Matt and me in the last few days before closing the skua shack for the winter. My mother had embroidered it on a patch and mailed it to me while I was working with Matt in Alaska because the island I was on was called "wind" in Aleut. He liked it so much that she sent him one too. Earlier in the season he'd mentioned to me that it would be cool as a tattoo. I'd drawn it on his wrist a couple of times so he could get used to the way it looked. But he wanted me to go first.

I'd learned to do stick and pokes from a crew mate and good friend on Midway Atoll. The first tattoo I ever gave on Midway was on her forearm: two red-tailed tropicbirds circling each other in a courtship flight. She imprinted a trail of albatross footprints on my foot, like the ones that cover the island, for the other beings we walk with, for the path we tread, and for the things we leave behind as we go.

After the first few tattoos were seen on people walking around, other people on the island approached me and asked for one. I ended up becoming the resident tattoo artist on the island, decorating skin with images of the birds we shared our lives with. Residents showed them off and visitors requested them as souvenirs. Eventually I had put together a whole kit and set it up with practiced fluidity, putting on rubber gloves and a headlamp, ripping sterilized needles from their wrappers. The Thais started calling me Dr. Nai.

I wove so many birds into skin, so many wings posed in flight—tropic birds, albatross, terns, a flock of frigate birds across an arm. I tattooed four out of five visiting air force firefighters. I tattooed people who were covered in tattoos. People who only had a few. People who'd never had one. People I was close to and people I barely knew.

Needle after needle, I wondered, What is it about the shapes of other species that moves us such that we want them embedded in our own body? What is it about the gestures of birds that so lends them to meaning? On an island surrounded by their feathered forms, tattoos ingrained their presence in our lives permanently. Were tattoos just a physical manifestation of all the ways we need other beings to be all of ourselves? As if skin were the canvas of our inwardness, and we were ever painting shape from feeling.

To Matt, who'd spent ten years of his life working with seabirds on remote islands, wind was the breath of life under a bird's wing. It is what they need to fly, it comes from the ocean and goes back to the ocean, just like the birds we study. To me, wind means something that is so strong and powerful because it is empty, it is made of nothing. It embodies change, it shifts the world as it passes, but if it ever stops moving, it dies. To both Matt and me, who had battled so much wind on this island, it was also the crazy joy of remote living and a mark of the world we shared.

This was probably Matt's last long field season, and the tattoo was like an inscription at the end of a book, commemorating his many years in the field. At the shack, we talked about what might come next, about how he craved a more stable life where he had time in one place to build something that felt rooted and continuous. He'd been living in Alaska, on a boat, house-sitting, in small apartments, ever picking up and moving when necessary, not in one place long enough to feel settled.

Matt, like me, was a bit of a restless soul, but we both felt this tension—a need for travel, for bare hills on the edge of the sea, for movement, and for a grounding return, to a place that is ours, to a life that is ours.

We talked about how we often felt that we had to choose between rootedness and fieldwork. A fieldwork lifestyle was analogous to travel and adventure, but it also meant that anytime I was in a job, I was also thinking about the next one—applying, searching, ever conscious of the need to have something lined up. It was exciting but also uncertain and insecure. I told Matt about the physician's assistant on Midway Atoll who was stationed as the medic on remote islands for part of the year, with the rest in her home state of Florida. From his own wilderness first-responder course, Matt loved learning about all the weird ways bodies worked and thinking about the puzzle of diagnosis. We talked about him pivoting to medicine, getting qualifications that would allow him to be gone for a portion of the year but still return to a stable life.

Matt imagined me as a journalist, writing articles on studies that uncovered new understandings of the ecosystems around us, traveling to remote spots like this one and interviewing researchers. As the child of journalists, I was wary of approaching the profession myself, even in ecology's orbit. I liked to think about what comes after studies are published—how to apply what we know toward building a more regenerative world, in which our society is integrated into the ecosystems that already envelop us. In which no person feels apart or above other species. In which our home biome feels as familiar to us as our own flesh and blood.

16.

Mid-March

The Cape slid closer to winter as our season drew to an end.

The fur seal pups were growing fast and getting ready to join the world of free and independent animals. The pups would wean at four months, closer to April. In March, mothers were still arriving onshore to feed the pups for a day or two, before taking off again to forage. To prepare for a future at sea, the pups were losing their baby fur (lanugo) and growing a fresh, warm, silvery coat. The young seals would be out in the ocean for three years before reaching breeding age. The territorial males were long gone from the beaches, back in the ocean recovering from the breeding season.

Fur seal puppies were trying out their swimming skills in the puppy ponds, pools of meltwater and rainwater that collected between hills down the center of the Cape. The ponds were much like kiddie pools, with all the same sounds you'd hear from a public pool full of kids—splashing, squealing, growling, more splashing—and like human kids, seal pups were a lot braver when they were with their friends. On my way back from the colonies, sometimes I took the inland route to pass by the ponds. I stood at the edge of the pools and the little seals all swam over and climbed out of the water to sniff me. They pressed their wet, whiskered noses against my pants and gnawed at my ski pole. Once they'd lost interest in the novelty of me, they splashed back into the ponds, learning to fly through water and leap into the air. Sometimes they'd float on their backs, fore-flippers crossed over their pale beige bellies, staring up at the sky.

It was hard to imagine these little squirts traveling hundreds of miles at sea in the next few months. The crew and I spent a whole evening watching a puppy that had taken up residence on the pile of wooden door covers in camp defend its territory. It flopped onto the door covers, trying to take up as much space as possible and pretending to relax, but jumped up, growling, to chase after any other puppies that approached. Then it would sprint back to the pile before other puppies took advantage of its absence. One of the other pups, meanwhile, kept getting distracted by whirling feathers. They got stuck in its whiskers, which it found confusing; eventually it ate a feather, smacked its mouth in distaste, and choked it out again. The whole ordeal had us giggling and crowding around the window, cocktails in hand. I couldn't decide who was more easily entertained, us or the puppies.

By the end of a season a whole camp culture would form, complete with its own mannerisms, references, inside jokes, and ways of being. At the end of my previous seasons the language I used with the crews had been shaped by months of common experiences. The Cape was no exception—we gave the animals little voices and developed personalities: how Weddell seals sing opera in their heads, how elephant seal brains move in slow motion, how chinstrap penguins slap anything that displeases them.

Moving around so much as a kid, ever the foreigner, I felt that I was looking at the cultures we landed in from behind a pane of glass. In international schools, others had landed there with me—dropped in an unfamiliar culture by their parents' jobs. Being together in an unknown world built an easy kinship, and the exploration of that world allowed it to bloom into the close and lasting friendships that are the only feeling of home I've ever known. Sometimes I wondered if I needed the raw novelty and challenge of an unfamiliar world—be it the bustling streets of a new city or the stark landscape of Antarctica—to feel close to the people who were there with me.

Ultimately, it was the other beings we shared the island with that brought us together, that shattered the glass and crumbled the walls. That gave us permission to be just as strange and chaotic as the world we lived in. The tethers between us were made of wet puppy noses, penguin flippers, skua squawks. When it came to the life out here—we were close only through them. The version of Sam I came to know excelled at impersonating seal habits and eccentricities. The Whitney I grew to depend on emitted optimism and cheer in even the most violent of blizzards. The Matt that accompanied my days was steady during captures, methodical in his data gathering, and endlessly amused by the penguins that surrounded us.

A maelstrom of two opposing feelings churned in my chest: I wanted to be done, to rest, but I also wanted to stay forever, anchor myself to this barren island and never leave. If the puppies and the penguins could slip past the edge of the only world they knew and swim off into uncertain vastness, then I could certainly summon the courage to step onto a Zodiac and be carted back to the world that was already familiar to me: a small house in the California woods.

The year I was born, my parents had bought a cabin in a national forest thirty minutes from Santa Barbara. While I was still in college, I returned to school after a field job, but after I graduated, I headed into this forest, where, under hundred-year-old oak trees, I processed the season and prepared for the next. The cabin was my in-between place—some intervals were just days, some months long. My aunts and uncles had houses nearby, and I wove through dry, scrubby sagebrush and across creeks to visit them. My aunts, uncles, and cousins plied me with food, heard my stories, then drove me into town so I could catch the next flight to my next field job. While I loved the cabin as a stopover, it didn't quite feel like my own. I was still happy to drift among duffel bags and airplane tickets.

After the season at the Cape, I would be returning to the cabin for a couple of weeks before heading out again. There were six months between the end of the first season and the beginning of my second season. With some cash building up in my bank account while I was on the island, I daydreamed a long itinerary of countries I planned to visit and pass through, aligned with a long list of globally scattered friends I wanted to see. It didn't hurt that traveling after a field job usually took some of the edge off the postseason slump.

Returning from my first few field jobs was rough, but over the years I'd gotten more used to the transition. In some ways, it got easier. Doug thought it got harder. The longer you spend in the field, the less you belong anywhere else.

After my first seasons in Maine and Alaska, when I stepped off islands and went back to a quaint college town for school, it was as if all the things I could feel changed with my surroundings. The raw dance of life and death is at its most dramatic on the islands I worked on: the tenderness of chicks, the rush of the breeding season, the struggle to stay alive through predation and weather, the final catharsis of fledge, the hope for survival. I felt so much of life's raw poignancy on islands, and it turbocharged my imagination and motivation. Heading back to school, I was always bursting with plans and ideas and expectations for everything I would achieve during the year. After St. Lazaria I started going on runs at 6:00 a.m. and dedicated myself to a long reading list of political theory. Back from St. George I had all these schemes for how I would turn our campus into an edible garden and eventually transform the college town into a food-forest heaven. Something about fieldwork always put me back in touch with myself and with the need to make the most of my life.

Usually within a few weeks of my return the spark would fade as life at school unfolded and I realized how much energy my island-born plans would take, how life is what happens when you're making other plans (as my grandfather used to say), and how nothing was as simple or as straightforward as I'd imagined, swept up in the rush and inspiration of the field season.

Back in a quaint college town, I felt numbed in comparison—without the wild churning of a marine ecosystem, the connections between other species and me were disparate and harder to grasp. It was somehow harder to feel, in general. Rather than wildflowers, I had ornamental trees; rather than nesting seabirds, I had sidewalk benches and cars. It was like slamming closed an existential door, like spreading flat concrete over all the textures of my heart. This is, for me, the center of the labyrinth. The basic, seemingly obvious idea that we need other forms of life, in all their expressive chaotic glory, to feel the full range of what a human can feel.

I have never been a particularly emotional person. In high school I was voted Most Grounded. In college, while dating someone distant, I took a silly online test for seeing if one was dating someone who was emotionally unavailable. I sent it to a friend, convinced I'd cracked the puzzle, and she asked me if I'd ever had the idea that, maybe, by chance, I was the one that fit these specifications. I read the test over again, and a sinking, dawning feeling washed over me. It was me. It has always been me. It was him too, of course, but I picked him for a reason, and it worked for a reason.

My emotive energy runs deep, like the lowest, darkest layer of the sea: the conveyer belt of cold waters, unseen, pulling along all the ocean's other currents. Islands, rising from the ocean floor, have always reached me there. Perhaps this is why, marooned in a largely dispassionate reality, the moments when I am entirely possessed by feeling evoke a particular fascination for me. I hold them in my mind, turning them over like an old toy, pulling them apart to better see the strings that make them move. That make me move.

Near the end of the season the whole crew fell to camp inventory, a daunting task. A *lot* of stuff is required to conduct science and sustain five people for five months on a frigid continent, and all of it must be accounted for at the end of the season. This kind of monotonous spreadsheet work could drive you crazy, especially after five months outside. Matt and I tackled all the food in the supply hut and droned on for two days: ". . . almonds, roasted, salted, three bags . . . almonds, roasted, unsalted, five bags . . . cashews, whole, raw, three bags . . ."; while Jesse and Doug did the toolshed: ". . . screws, galvanized, two inches, hundreds . . . screws, galvanized, one inch, hundreds . . ."; and Sam and Whitney picked away at the fur seal lab: "syringes, ten milliliters, twenty-four . . . syringes, twenty milliliters, one hundred and four . . ."

After Matt and I finished the supply hut, Sam and Whitney were still counting screws and syringes. Matt and I took on main hut and the crawl space between the ceiling and the roof, packed full of old, old stuff. In the crawl space, I had to drag myself along on my belly through two narrow passageways and squirm to squeeze under the rafters. Then I rummaged and rearranged until I was totally boxed in, yelling over the pile to Matt, who was standing on a ladder with his torso poking through the attic opening in the ceiling with a laptop, and who was already cranky before we began all this.

"There's a black bag here that says *gloves!*" I hollered over to him.

"Where are you!"

"Hang on." The sound of shuffling. Each narrow passage through the rafters was labeled with a number and orientation. "I'm in Ten West!"

"What kind of gloves!"

"Rubber, I think! Do you want me to count them!"

"Yes!"

"Okay, one second! Medium, five—large, ten—small, twelve . . . ," and so on.

By the end of the day, I felt brain-dead. At dinner we needed to discharge all the pent-up energy, so while polishing off the rest of the wine, all the frizzy, goofy humanity reared its head after hours and hours of dullness. We were still stuck in inventory language: "Can you pass me the wine, comma, red, comma, delicious? Where is the salt, comma, sea?"

Jesse's rants against clutter grew longer and increasingly chaotic. Doug's movie quotes became more obscure and he stopped bothering to ask if we'd seen the movie they came from. Matt had cleaned the Chilean hut and moved back into main hut, moving quietly among the rest of the crew and disappearing early into his bunk. At night Sam and I talked in whispers while cleaning up and getting ready for bed: How's it going with Matt/Whitney? Do you think you're ready for next year? How are you feeling about everything?

One night in early March, I knelt in my bunk, wrestling with my stuff and trying to figure out how many clean pairs of socks I had left. My giant bag of socks was stuffed into a milk crate nailed to the wall above the foot of my bed. While rummaging, I stumbled upon the bottle of pinot noir that I had buried there back in October. I'd bought it when I went wine tasting with an old friend who visited me in California right before I flew south for the season. After many months I felt disconnected—as if the world out there might as well not exist and my previous life might never have happened. But here was a bottle of wine from warmer places, solid in my hand, a real object that came from a real place that I had been to and that still existed, and I felt as if it had been years since I'd walked through a small town on a summer evening. I pulled the bottle out for the night—there was no special occasion, save that of having a delicious bottle of wine to drink.

In college, two friends and I ran a wine club, using school funds to buy fancy bottles and hold wine tastings on Monday nights. We demanded that everyone attend in semiformal attire, and we sat at the front of the room, presenting our wines with a flourish, giving everyone a splash each, telling them what flavors to look for. We weren't sommeliers, and the events were more performance art than culinary education, but they were a wonderful alternative to a drinking culture defined by shots and shotgunning cheap beer. The Sunday sign-up sheet always filled up within minutes of being sent out.

With the same dramatic flourish from my wine club days, I presented the wine bottle to the crew and we uncorked it before dinner. There was enough for each of us to have a small glass, and I sat with the wine in my mouth, breathing air over it and absorbing the flavors, closing my eyes and letting it trickle down my throat. I could not believe how good it was. We had been surviving off boxed wine for so long I'd forgotten wine could be anything other than basic, bitter, and flat. My senses were dulled by the cold, the dampness, the canned and frozen food, the musk of my own unwashed body. The wine pierced through the haze and danced on my tongue, like a herald of warmer places. Whitney closed her eyes and made happy humming noises. She had a keen palate; she'd worked at a distillery and other food joints and could always tell what ingredients were in other people's concoctions. When dinner was ready, Whitney and I were still nursing our glasses as if these were the last cups of wine we'd ever drink. At the table Whitney pushed away her plate, unable to bear the thought of flavor contamination.

That night, after dinner, I wandered outside, stepping off the deck onto antarctic soil and looking out at the water. I peered at the moon, glowing behind clouds. The snow fell lightly, the breeze as sharp as ever. Turning back, I saw a square of golden light: the window of main hut, the only light in all this darkness, which I knew led to warmth, food, and people. I saw the crew talking and laughing, the Christmas lights glinting behind them.

It was the strangest feeling to see everything in the whole world that sustained me reduced to a floating square of yellow. My survival depended on those four thin walls and a few rusty propane tanks. I blocked out the window with an outstretched hand, and suddenly there was nothing, only wind and water and death. I felt the isolation of this small human outpost, the coziness of my fragile home, and the privilege of staring into the undisturbed darkness that engulfed it.

I thought about the scientific incentive that built all of this, the decades of dedication, the journal articles that were born within that square of yellow and sent out into the world. The clear structure of the scientific method is like the grammatical rules of a language, organizing information to yield meaning. Science seemed like a kind of love language, coming from the depths of a heart as well as a keen mind. But it was not *my* love language.

From my first field season and first biology textbook, I have been drawn toward the ways our understanding of ecosystems informs our understanding of ourselves and of our society, the way in which biology bleeds into culture, ecology into art. In the field, I tried to see through the eyes of Emerson or Thoreau, through the eyes of Tlingit, Aleut, and Hawaiian storytellers to unravel the way Western culture conceptualizes ecosystems and people's place in them.

Antarctica seemed like unique ground in this respect. Learning about the history of the continent during the season through the tomes in the book tote often made me think about the continent's cultural presence, as I sought to contextualize my own experience. I thought about the way Antarctica is alive in Gothic novels, in paintings, the way it acts as a fulcrum in our collective psyche, representing all things distant and nonhuman. The way the first explorers came and wove the continent into the human conception of the world. And how: the story of an expedition, a painting of inky scenery, the symbolic landscape of our collective psyche. I considered the accumulation of stories told about this far and foreign place, and the way art leads to conversations—which lead to culture and identity, which lead to values, political will, policy, and protection.

I still didn't know exactly where this interest would lead me, only that my days in the field were numbered. I wouldn't be collecting data forever. I didn't have a plan, exactly, for what would come after my second season, only the thought that I'd like to be close to science, if not within it. Maybe next season, spurred by the introspection of isolation, I'd ride a wave of insight and the shape of my life's calling would click into place. I have time, I thought, although the need to latch on to something lasting often stirred in me.

———————

We worked hard to stay a bit ahead of schedule in closing camp so we could have the last day to ourselves. The morning promised a breezy, sunny day, and we made our plans. Whitney, Sam, Doug, and Jesse hiked off to Punta Oeste because they didn't get to come the second time Matt and I went, and it's a beautiful, remote spot. Matt, faced with the choice of a group of people versus no people, opted out. He wanted to hike north to the penguin colonies, where we'd spent so much time. I went with him. We walked along in silence, absorbed in our own thoughts, traveling a familiar path with a familiar person for the last time.

The colonies were much the same as they were the week before. Quiet, molting chinstrap penguins. Gentoo chicks, fully feathered, running around. The fledged chinstrap chicks were off in the ocean, traveling north of where the pack ice would form in the winter.

The skua shack was winterized and boarded up. Walking past the shack's peeling walls, Matt spotted a penguin on the beach that looked a bit bigger than the others, and we wandered down to find a king penguin standing upright on the rocks, looking around, sun alighting on the bright yellow whirl of feathers at its nape. The king penguin was a yard tall and barrel-chested, towering over the gentoos and chinstraps that passed by. King penguins breed largely on subantarctic islands, the closest colony to the Cape being in South Georgia. Sometimes they wandered the Southern Ocean and showed up farther south, on our shores. A subadult fur seal male was harassing this king, lunging, snarling, and the king snapped at the seal by extending its snaky neck out farther than I thought physically possible, opening his long, tapered bill menacingly. The fur seal retreated, and the penguin retracted its neck and burrowed its head into its broad shoulders. It peered at us with round, black eyes. The king's massive flippers, poised by its barrel chest, seemed to go on forever. We wondered what it would be like to be slapped by those flippers.

Matt tried to process that he'd probably never return to Antarctica. "I wonder when I'll get to see a chinnie again," he said, staring at the familiar forms of our study species, clustered on the shit-stained rocks, shedding feathers like water droplets.

"Probably in a zoo," I suggested unhelpfully. "It's not the same, is it? Not like this."

"Nah," he said, peering out.

I was spared the goodbye—I knew I'd be back next season to take his place as seabird lead, and to pass penguin duties on to the next penguin technician. Matt and I sat down on a rock near the colonies and soaked it in. Then we hiked back to main camp together for the last time.

The ship was scheduled to come in the morning. The only things left unpacked and unwinterized were those things we'd need to sleep and get going in camp: the coffeepot, indispensable; our bedding; the chairs; and two cold pizzas. Most of the kitchen was bagged, most of the electronics had been uninstalled and stashed in the electronics clamshell, and all our personal stuff was packed. All that was left to do in the morning was to pack our bedding, bag our mattresses, put on our final set of gear, clean a few dishes, and board everything up.

The ship appeared on the horizon at noon the next day. Soon, it hailed us on the radio, the last electronic device to be kept running. By 1:00 p.m., the orange inflatable boats approached our shores. Whitney, Doug, and I stayed in main camp to finish up, while Sam, Jesse, and Matt headed down to the beach to help load our stuff.

One of the last things that must be done in camp is to switch out the four-wheeler's wheels for snow treads, so it can be driven straight out of the shed the next year to haul gear at camp opening. After the last load went down to the beach, Whitney drove the four-wheeler up onto the deck on two rickety ramps. Like so many other things, Whitney and Matt had only switched to the treads once last season, and they tried to remember as Sam and I tried to learn. The four of us puzzled over bolts and ball joints, wrenches in hand, our season's gear disappearing from the beach like sand through fingers, the last few minutes in camp ticking away. Treads finally attached, Whitney backed the UTV into the tiny shed, an inch of space on either side of the vehicle.

I headed back into main hut to take a final look around before we boarded it up. Everything that could be bagged in plastic was, to protect it from mold. Wires stuck out of the walls, covered in electrical tape, like vines probing the air. The floor was bare and clean. The bunks were stripped and the mattresses propped up and bagged. The whole kitchen was bagged. The space barely looked habitable, just like what I'd found at opening, when this was all new to me.

I scanned the space for anything else that needed doing before we headed off. Suddenly my perspective shifted, and instead I saw all the accumulated hours of my life spent between these walls. Sam walked in, saw me looking, and wrapped me in a bear hug, and we squeezed each other for everything we'd lived and everything we'd be leaving behind. I thought back to the first day we landed in Antarctica and saw the gentoos trailing by camp. "Look! A penguin!" we'd exclaimed, near strangers, wide-eyed with awe. Outside, a handful of gentoos walked north on the penguin trail between camp and El Condor. Just as the penguins had on our first day, as they'd done the whole time I'd been here to witness it, as they had for decades prior and hopefully would for decades more.

———————————

Riding off on the Zodiac an hour later, I pinned my eyes on camp until it receded completely into the mist. Only then could I turn my head and look forward to the approaching ship, a looming wall of orange, swaying on turbulent waters. We took turns climbing up the side of the ship on a rope ladder, timing our first steps with the peaks of the ocean swells. The ship deck clamored with metallic sounds, fraying my nerves. When I saw Jesse's familiar, competent face amid the chaos, my agitation abated.

Standing next to all these people with clean clothes and faces made me hyperaware of how scruffy and filthy I was. My bib wore the accumulated grime of five months in camp, my fingers were crusted with four-wheeler grease, my face was oily from activity, my hair was one big knot secured by three ancient hair ties, made half-blond with leftover puppy bleach, my feet smelled like death beneath two layers of stiff socks. The boys looked like castaways with their grizzled beards.

After dumping my personal bag into the washing machine, I stepped outside to watch the ship travel around the island. We circled the west side and turned south. I could see Livingston's distant mountains from a different angle, Cape Shirreff now reduced to just a jutting bit of rock. It felt so strange to see this land from the water. As if I were floating outside my own body. Matt pulled out his binoculars and scanned the waters for penguins. I peered into the mist and tried to spot the small square splotches that were the huts we'd lived in, but we were too far away already for such details, the mountains looming closer now than the Cape. We traveled west, and I saw the ice-capped hills of the interior and the rocky, snowy Byers Peninsula, the other main snow-free part of the island. Then we curved north toward Chile, the distance growing between the ship and Livingston Island, the South Shetlands, the northernmost crumbs of the antarctic.

We slowly left the Antarctic Peninsula behind, which curves south like the knobbly spine of a seahorse, splitting southern waters into the Bellingshausen Sea and the Weddell Sea. The Larsen Ice Shelf hugs the peninsula's curve, extending from the eastern coast. Deeper south are the land masses of East and West Antarctica, divided by the Transantarctic Mountains. These remote peaks tower over ice, snow, and the few settlements humans have built to peer into the frozen land and discern its secrets.

Being on the ship was an in-between time, where I had no responsi-bilities and could let the experience wash over me. I thought a lot about the penguins out there zooming through watery depths, hunting small crustaceans. Immersed in a world I could not even imagine, a world that receded as we puttered north. Chinstraps would live on open ocean until the adults returned to the Cape again in October for the breed-ing season, when I would meet them back on the shores of Livingston Island. The Cape's gentoos, never far from home, would live through winter's dark days in the South Shetlands, hunting for fish and krill.

The data I collected over the season would be another notch in the long-term data set, charting the ecosystem's changes through its top predators. Even together with the Cape's many decades of monitoring data, our observations of the dynamics at the Cape were a window into just one specific region of Antarctica, to be interpreted in context with other regions. Jefferson, Doug, and Mike would do just this, along with other scientists, when they brought their data before the scientific com-mittee of CCAMLR.

While CCAMLR was designed to regulate fisheries in the Southern Ocean, Mike has told me that he thinks the impacts of the krill fishery are minimal compared to the impacts of climate change. Though the effects of climate change on the Western Antarctic Peninsula are stark and dramatic, the effects on the rest of the continent are less clear. In some areas, ice is increasing rather than diminishing. On the east side of the Antarctic Peninsula, the ice is holding steady in the notoriously icy Weddell Sea, and so are the area's populations of Adélies. Chinstraps do not nest east of the Antarctic Peninsula—too icy for them. There is still much to be studied and understood about the Southern Ocean's complex response to climate change.

Living in and learning about Antarctica forced me to see beyond the bounds of my own life to the continent's long, slow rhythms. I found my sense of time elongated in polar lands: the events of a year meant nothing on their own. A season's dynamics could only be understood alongside decades of data and the cyclical rhythms of the continent itself, which occurred on many time scales at once. Pressure fronts that shift every few years influence weather across the continent. Krill populations tend to peak every five years, when the conditions are just right for their reproductive cycle. Annually, the ice forms and melts, a pattern that all marine life in southern latitudes have adapted to and grown to depend on. Seasonally, species breed on land and return to the sea. Daily the sun orbits the continent's windswept landscape. And every minute that passes, the waves lap onto dark rocky shores, rising and falling with the tide.

Time wasn't the only dimension that felt stretched taut by the bounds of my perception. On the top of a hill on a clear day, I could see the island laid out before me, bare of vegetation, with the ocean stretching far into the distance. The only thing that limited my view was the very curvature of the earth. From up there, the land felt huge and ancient, steady and patient, dwarfing the whirling machinations of my own mind, and stilling them.

Unique among continents for a million reasons, Antarctica forces us to expand our frameworks and push beyond our usual perspectives. Fringed with teeming waters, the continent has already inspired a new, holistic management approach to regulating fisheries. CCAMLR is tasked with understanding an ecosystem through the relationships species have with one another and with their environment. As a result, the Southern Ocean is considered one of the best-managed fisheries in the world.

The Antarctic Treaty system is a unique example of international collaboration toward conservation ends. CCAMLR is not without its challenges, but it is also not without its strengths: I still find it amazing that, even in a globalized culture steeped in the ethos of profit and exploitation, an international governance structure exists to protect and honor an ecosystem, in all its complicated wholeness, covering a massive region of this earth. CCAMLR has worked to protect some of the Southern Ocean's most sensitive ecosystems, and it can do so again. I hoped that the regulations for the Marine Protected Area around the Antarctic Peninsula would soon be passed.

A lot can be learned in our efforts to mitigate climate change by looking at Antarctica. Climate change is a problem that requires the same long view, spanning far into the past and far into the future, beyond the bounds of our own lives. It requires us to examine closely all the interconnected pieces of our earth's ecosystem, including the systems we have built. And like Antarctica, climate change forces us to think beyond our constructs of state and sovereignty, to an integrated view: ourselves as one species of many and our home as a living, breathing, feeling Earth.

———————

On March 20, I fell asleep at sea and woke up tethered to a continent.

I'd made it.

Stepping outside onto the dock was like passing through a portal. People were roping up ships, hauling gear, walking—people, people, people. Sam, Whitney, Matt, and I walked down the dock out to the sidewalk, while Jesse and Doug were off attending to other matters. I could not believe the trees after so many leafless months. Wind-worn and ragged, they grew like a revelation from dirt patches in the pavement. We moved down the sidewalk wide-eyed and close together, remembering and readjusting to cement and cars and faces of unknown people. I looked out warily into the world I was about to reenter, timid with strangers, attached to my crew. Looking for echoes of the world I'd just left in the one I used to know.

Any field season meant total immersion in our study sites, and that had never been truer than at Cape Shirreff. My job was to monitor, literally, to observe, and I was on the island for the whole breeding cycle, from pebbles to eggs to chicks to fledgers to silence and the whirling of shed feathers. For a few months the study site was everything. I couldn't go anywhere else, I couldn't reach beyond the beach, so I went deep instead, sought a profound intimacy with the place in which I lived.

But at some point all field seasons come to an end. A boat arrived and we all clambered into it, half of us leaving the island for the last time.

I left the Cape with the lasting impression of a stark and stoic landscape. I also left with skills that I suspected would never come in handy again (except for the next season): how to massage the krill from a penguin neck, the delicate art of dumping a bucketful of poop into the ocean, how to pluck a whisker from a wriggling twenty-four-hour-old fur seal puppy, how to maneuver a leopard seal into a tarp, the mind games needed to catch a growing skua chick, and the amount of force necessary to loosen a frozen wing nut. I was looking forward to next season and to leading the penguin research. I hoped we'd have a good new cohort of technicians, with the particular mix of personality traits that is so common among wildlife field-workers: a goofy sense of humor and a propensity not to take oneself seriously, an iron will, a huge tolerance for discomfort, a steady and grounded sense of self, adaptability, creativity, and, above all, a love for the job.

As my plane departed Punta Arenas, I stared down into the seemingly infinite expanse of the ocean, and a remote ecosystem on the other side of it weighed heavy on my heart. My grief was a dark squall, looming on inevitable winds. I could not control or escape it. But like so many Cape blizzards, it would pass. All there was to do was put my head down and get through it. I would get wet. But I would also be dry again.

My body molded to the familiar shape of an airplane seat, which also felt a bit like home. The waters stretched below me, subdued, rippling light. I had a lot of travels planned, I would be moving a lot, and quickly, for the next six months, before I was due back in Punta Arenas in October for my second season. There was so much to look forward to. I closed my eyes and leaned my head against the window as the plane hurtled through the sky. Onward.

EPILOGUE

For the six months in between field seasons, I traveled. I wanted to go everywhere and do everything. After leaving Punta Arenas, I hung out with Renato in Santiago, Chile, for a few days, as we'd planned to do while we were at the Cape, then met up with my sibling in Buenos Aires, Argentina. After a week of reminiscing in our childhood haunts, I went back to Chile and met Matt in San Pedro de Atacama. We traveled together for a couple weeks through the desert and up into Bolivia. In this landlocked country, we still managed to find an island in the middle of Lake Titicaca. I flew back to California to regroup and see my family, but before long I was off again: Hawaii, Malaysia, Thailand, Laos, Singapore. I went to India, where I stayed at an ashram and studied philosophy for a month with two friends from college. I went to Turkey, crossing the Bosporus in big white boats and snacking on crunchy, tart street food. I went to Spain, where I saw my cousins and my grandmother and walked across the country on the Camino de Santiago. Then New York, where my parents were currently living. I circled back to California, and before I knew it, I was headed south to Chile to start my second season.

The season was different, as I expected it to be. Sam and I were in charge, trying to teach while we tried to remember, passing on our tips to the next cohort. There were new people, new technologies, changing protocols. I sensed that big changes were coming. We did more camera work and drone work. With drones, we could count populations without having to physically walk through the colonies and section them out with ropes. Doug and Jefferson did a study comparing our usual counting methods with the drone method and found that drones caused minimal disturbance in comparison. They measured disturbance by setting up audio recorders while I and the new seabird technician did our manual survey, comparing the penguin noise we caused in the colonies to penguin noise caused by overhead drones. This shift in technology was one of many that eliminate long and arduous tasks that had characterized penguin monitoring. I knew that it was better for us to get the information we needed in a way that caused less disturbance, but I couldn't help but feel sad for the technicians who would come in the years afterward and never experience laying bright rope between chinstraps and counting a dozen subpieces while birds tugged the rope and wandered over boot-clad feet.

The Trump administration tightened NOAA's budgets, and Jefferson and Doug explored more ways of getting the same information in a more efficient manner and with less disturbance. These increased uses of technology could not have come too soon, since these budget cuts also meant that the season had to be shortened by two months. The last season to run for five months was also my last, in 2017–18. In 2018–19 and 2019–20, camp ran only from December to March, just in time for perinatal captures and chick-rearing device deployments. Cameras set up the previous season caught enough of the laying and hatching to keep the data set continuous. The protocol for diet sampling was no longer approved by the animal care committee, being deemed far too invasive. Jefferson has been figuring out other ways of getting diets, including waiting until the adult is about to regurgitate and then quickly moving the chick to let the krill hit the ground and be collected. Oh, and alcohol was banned at NOAA field camps. Good thing the season was only three months long.

The structure of the crews has also changed. With the cameras collecting data, there is less need for two dedicated people for seabirds and two for pinnipeds. Instead, three people are hired year-round—three months in the field, and the rest of the time in San Diego working for the program, on a three-year contract. The program started making the shift in 2019, finalizing the hire of the new technicians in 2020. Then the COVID-19 pandemic hit, and the field season was canceled due to the logistical nightmare of international travel and quarantine needs. Fortunately, people from the National Science Foundation, headed to the permanently staffed Palmer Station, were able to go onshore and replace the camera batteries and memory cards in the penguin colonies and the fur seal rookeries. This enabled just the bare bones of the data set to be continued. All over the world, these long-standing ecological-monitoring data sets were affected by the pandemic. In 2022, the program managed to send a team down for just a few weeks after almost as long in quarantine in California and Chile. Their season was a mad rush of device deployments and population counts. Sam works for the program now as a data scientist, and he and Doug were the only people on the team that had been at the Cape before.

In many ways I experienced the last of the field seasons as they'd been since the conception of the program. This is particularly true for the huts I called home—NOAA's Antarctic Ecosystem Research Division is investing in a new camp, to be built in 2022. The old one will be dismantled. I don't think I'll ever go down there again, but it still feels bittersweet. I'm excited for future field techs to live somewhere a little more functional, though I can't help but mourn our long-suffering plywood walls. I feel that the hut is so heavy with history it should be airlifted to a museum.

In CCAMLR's 2021 meeting, held online, the three Marine Protected Areas previously proposed—one off the coast of East Antarctica, one in the Weddell Sea, and one around the Antarctic Peninsula—again came to a vote. It was CCAMLR's fortieth meeting, and the sixtieth anniversary of the Antarctic Treaty system. More countries signed on to the proposed MPAs, but they still failed to pass without China's and Russia's support.

Mike retired and is always off on a new adventure. To no one's surprise, Whitney got into multiple vet schools and is in the middle of her studies. Matt is a medical technician in a small town in Alaska, living in a cabin up on a hill in the middle of a wildlife reserve, where bears wander by under the balcony and moose chomp on berries nearby. Doug and Jefferson still run the program, with Sam as a permanent part of their team. Renato went on to work as a science coordinator for the Chilean Antarctic Institute, herding scientists around to their study sites on King George Island.

As for me, I write this from Wellington, New Zealand, where I have made my home for now. I did a few more field jobs after my allotted two seasons at the Cape—I wasn't quite ready to leave fieldwork behind. I worked on an island off the coast of California, on a tiny mountain island in American Samoa, and on the North Slope of Alaska before applying to grad school and sleeping on the futon in Matt's cabin for six months, working in the town bakery and watching summer turn into winter. I needed some time in society to prepare to move to New Zealand to complete my master's in conservation biology. This writing project was a grounding force through all of those changes.

In my second season at the Cape, Jefferson, the seabird lead researcher, stayed with us in the latter half of the season. He brought with him two small cameras to deploy on penguins. Finally, technology had advanced to the point that video loggers were small enough to do this. We chose a beefy chinstrap female with a four-week-old chick for a camera deployment. I attached the camera the way I'd attached the other devices, right in the middle of its back, over its spine. The loggers could only capture eight hours of video, so it would be just a twenty-four-hour deployment.

I'd suggested we attach small time-depth recorders as well, so we could match ocean depths with what we saw in the video. When we showed up in the colonies the next day and saw the penguin by her nest, camera intact, the new penguin tech and I almost fell over in our rush to put on our penguin clothes. We were so excited when we got the camera back. It was the first time video loggers were deployed on penguins at Cape Shirreff. That afternoon, Jefferson hurriedly hooked the thing up to a computer and downloaded the clips. It is hard to describe the thrill of retrieving a camera and plugging it in, with no idea of what you are about to see but with the certainty that no one on this earth has seen it before.

When Jeff opened the first video, we saw a view of the roiling ocean, and the back of the penguin's head in the foreground, floating on the surface, beads of salt water on its nape. In a moment it dove into the ocean, swimming just below the surface, then other chinstraps appeared on the screen, flying through the water, their forms from behind like pointed balls, sleek and aerodynamic. They were commuting from the colonies to their foraging grounds. I hadn't known that chinstraps commuted to the krill swarms and hunted together.

Once they arrived at some undetermined point in the ocean, known only to them, the penguin began diving into the deep blue. Suddenly— krill. Swarms of krill suspended in the water, the penguin weaving through them, whipping its head around to grab krill as it went. We saw other chinstraps zooming through the krill swarms. When the penguin emerged from a dive, it bobbed on the surface, shaking water from its head, reaching toward its back to preen. We could see the same sky the penguin saw, the same ocean, the same swarms of krill. We even spotted a banded penguin it was foraging with. In the evening we ate an apple pie while watching the penguin footage on the projector, rum cocktails in hand, cheering every time the penguin caught some krill.

We could get so much more information from video: that penguins foraged together, how successful their dives were, how many krill they consumed a minute, and the density of the krill swarms. Scientifically, the camera brought so many possibilities to better understand the chinstraps' foraging patterns during the breeding season.

But it was more than that. It is hard to describe how magical this experience was, after orbiting their world for so long. After so many months of knowing them only on land and guessing what their trips at sea were like. The camera footage let me watch the view from the chinstrap's back for hours, let me know the rhythm of its hunt and the pattern water made streaming off its waterproof head. I even watched its chick bother it when it returned to the nest. It was as if I went with the penguins to hunt krill in frigid waters. As if I were *there*, zooming through the dark blue waters with other chinstraps, snapping up mouthfuls of krill, resting on the surface under a vast, cloudy sky, the sun casting a yellow pallor on the horizon as it set.

I remember watching and feeling in waves a familiar awe: that this world is also my world.

ACKNOWLEDGMENTS

M y deepest gratitude to:

The crew I shared my time with at Cape Shirreff, in particular those who feature in this book. Thank you for being such a joy to be with and to write about. Douglas Krause, Michael Goebel, and Jefferson Hinke, for steering the program with such heart and intelligence, and the whole NOAA Antarctic Ecosystem Research Division for the invaluable work you do.

Matt, for everything you taught me and for letting me crash with you while I figured out my life. I can't imagine any of this without you. Sam, for holding it down with me for two seasons, for digging up data for me, for the consistent stream of seal memes, and for being such a wise and thoughtful friend.

Nina Karnovsky, for the joy with which you teach and work and for opening the door of fieldwork to me. Susana Chávez-Silverman, for being the first to make me believe I could actually be a writer.

Lucy V. Cleland, my wonderful agent, for seeing what this book could be in the mess of pages I first sent her all those years ago and for shepherding me through the process with such grace. Sarah Gold-

Acknowledgments

berg, for taking on this project, and Sally Howe, for deftly shaping the book with wisdom and curiosity. Everyone at Scribner who had a hand in transforming my manuscript into this book. I couldn't have asked for a better team.

My best friend, Gabi, for being with me at every step of this very long process and for the honest and insightful writing advice that has guided my hand more times than I can count. Kai, for his absolute and unwavering faith in me. I have leaned on it when my own faith has faltered. Keely, for the first tattoos and for being a part of this story.

Mauricio, por calmar mi ansieded en los momentos más estresantes y por ser una fuente de fuerza e inspiración (y perreo intenso). Mis brujitas Isa y Maite, por las celebraciones y por quererme como soy. Kevin, for patiently explaining antarctic atmospheric pressure fronts to me, for the hot chocolate chats, and for hauling me out to the mountains when I needed it the most. The crew at Chilka St. for nourishing me and making space for big feelings in the last few stages of the process.

So many more friends have helped me through these years of writing than those I've named here. You know who you are—thank you.

The Love clan, for being such a wonderfully loving and chaotic family and for cheering me on. For always taking me in whenever I showed up and seeing me off again when I left. Suzy, for the introduction that started everything and for believing I could do it.

Finally, my beautiful parents, for teaching me to write and giving me a life worth writing about.

NOTES

Prologue

xiii *decreasing the number of annual days of ice and the amount of ice overall:* Hugh W. Ducklow et al., "West Antarctic Peninsula: An Ice-Dependent Coastal Marine Ecosystem in Transition," *Oceanography* 26, no. 3 (2013): 190–203.

Chapter I: Mid-October

4 *88 percent of which is covered by an ice cap:* Lyubomir Ivanov, "General Geography and History of Livingston Island," in *Bulgarian Antarctic Research: A Synthesis*, ed. Khristo Pimpirev and N. Chipev (Sofia: St. Kliment Ohridski University Press, 2015), 17–28.

6 *Vast swarms spanning up to twelve miles:* Stephen Nicol, *The Curious Life of Krill: A Conservation Story from the Bottom of the World*

(Washington, DC: Island Press, 2018), doi:10.5822/978-1-61091
-854-1.

6 *holding the whole ecosystem together:* Andrew Bakun, "Wasp-
Waist Populations and Marine Ecosystem Dynamics: Navigating
the 'Predator Pit' Topographies," *Progress in Oceanography* 68,
nos. 2–4 (2006): 271.

6 *directly or indirectly on krill for their survival:* George M. Watters,
Jefferson T. Hinke, and Christian S. Reiss, "Long-Term Observa-
tions from Antarctica Demonstrate That Mismatched Scales of
Fisheries Management and Predator-Prey Interaction Lead to
Erroneous Conclusions About Precaution," *Scientific Reports* 10,
no. 1 (February 11, 2020): 2314.

Chapter 2: Late October

18 *moving west along the continental shelf:* Malgorzata Korczak-
Abshire et al., "Coastal Regions of the Northern Antarctic Peninsula Are Key for Gentoo Populations," *Biology Letters* 17, no. 1
(2021), https://doi.org/10.1098/rsbl.2020.0708.

24 *on the vessel* Te Ivi o Atea *in the seventh century:* Priscilla M. Wehi
et al., "A Short Scan of Māori Journeys to Antarctica," *Journal of
the Royal Society of New Zealand* (2021), 1–12, https://doi.org/10
.1080/03036758.2021.1917633.

27 *The prospect of years at sea . . . in order to man the ships:* Ben Maddison, *Class and Colonialism in Antarctic Exploration, 1750–1920*
(London: Pickering & Chatto, 2014).

Chapter 3: Early November

31 *have been strongly linked with the extent of sea ice the previous winter:* Angus Atkinson et al., "Long-Term Decline in Krill Stock and Increase in Salps Within the Southern Ocean," *Nature* 432, no. 7013 (2004): 100; Langdon B. Quetin et al., "Ecological Responses of Antarctic Krill to Environmental Variability: Can We Predict the Future?," *Antarctic Science* 19, no. 2 (2007): 253–66.

31 *sea ice has decreased in duration by almost one hundred days:* Ducklow et al., "West Antarctic Peninsula," 190–203.

31 *and in extent by 47 percent since 1979:* Jaume Forcada et al., "Responses of Antarctic Pack-Ice Seals to Environmental Change and Increasing Krill Fishing," *Biological Conservation* 149, no. 1 (2012): 40–50.

31 *a 12 percent reduction in phytoplankton:* Martin Montes-Hugo et al., "Recent Changes in Phytoplankton Communities Associated with Rapid Regional Climate Change Along the Western Antarctic Peninsula," *Science* 323, no. 5920 (2009): 1470–73.

31 *seals, penguins, and whales:* Ducklow et al., "West Antarctic Peninsula," 190–203.

35 *1.7 million antarctic fur seals were killed:* Ivanov, "General Geography," 17–28.

36 *1.8 million whales were killed:* Lisa T. Ballance et al., "The Removal of Large Whales from the Southern Ocean: Evidence for

Long-Term Ecosystem Effects?," in *Whales, Whaling, and Ocean Ecosystems*, ed. Phillip J. Brownell Jr. et al. (Berkeley: University of California Press, 2019), 215–30.

37 *back to 93 percent of their pre-exploitation levels:* Alexandre N. Zerbini et al., "Assessing the Recovery of an Antarctic Predator from Historical Exploitation," *Royal Society Open Science* 6, no. 10 (2019), https://doi.org/10.1098/rsos.190368.

37 *not even reached one thousand:* Douglas J. Krause, Michael E. Goebel, and Carolyn M. Kurle, "Leopard Seal Diets in a Rapidly Warming Polar Region Vary by Year, Season, Sex, and Body Size," *BMC Ecology* 20, no. 1 (2020): 1–32.

Chapter 4: Mid-November

54 *as a sublime landscape where one could encounter God:* William Cronon, "The Trouble with Wilderness; or, Getting Back to the Wrong Nature," *Environmental History* 1, no. 1 (1996): 7–28.

55 *"You find thus in the very sands . . . and are pregnant with it":* Henry David Thoreau, *Walden; or, Life in the Woods* (London: J. M. Dent, 1908).

57 *acquired the rootedness bestowed by local ancestors:* Elizabeth Leane, *Antarctica in Fiction: Imaginative Narratives of the Far South* (New York: Cambridge University Press, 2012), 1–21.

Chapter 6: Early December

75 *gas anesthesia methods:* Nicholas J. Gales and Robert H. Mattlin, "Fast, Safe, Field-Portable Gas Anesthesia for Otariids," *Marine Mammal Science* 14, no. 2 (1998): 355–61.

Chapter 7: Mid-December

91 *Trip duration, which ranged from three to forty-eight hours:* Rory P. Wilson and Gerrit Peters, "Foraging Behaviour of the Chinstrap Penguin *Pygoscelis antarctica* at Ardley Island, Antarctica," *Marine Ornithology* 27 (1999): 85–95.

92 *they can only tolerate ocean temperatures up to 5˚C:* Nicol, *Curious Life of Krill.*

92 *some evidence that with warming temperatures the range of krill is contracting southward:*

  • Atkinson et al., "Long-Term Decline in Krill," 100.

  • S. Moorthi, "Krill vs. Salps: Dominance Shift from Krill to Salps Is Associated with Higher Dissolved N:P Ratios," *Scientific Reports* (Nature Publisher Group) 10, no. 1 (2020).

  • A. Atkinson et al., "Krill (*Euphausia superba*) Distribution Contracts Southward During Rapid Regional Warming," *Nature Climate Change* 9, no. 2 (2019): 142–47.

92 *models predict that krill populations could largely shift south:* A. Piñones and A. V. Fedorov, "Projected Changes of Antarctic Krill Habitat by the End of the 21st Century," *Geophysical Research Letters* 43, no. 16 (2016): 8580–89.

Chapter 8: Late December

105 *Sitting on the bottom of the globe, Antarctica is described:* Leane, *Antarctica in Fiction.*

Chapter 9: Early January

123 *continually feeding and expelling waste:* Atkinson et al., "Krill (*Euphausia superba*) Distribution," 142–47.

123 *about 23 million tons of carbon dioxide:* Barbara Cvrkel, "A Network of Marine Protected Areas in the Southern Ocean," Pew Charitable Trusts, 2019.

Chapter 10: Mid-January

129 *They are equipped with sharp, curved canines:* G. L. Kooyman, "Leopard Seal (*Hydrurga leptonyx* de Blainville, 1820)," in *Handbook of Marine Mammals*, ed. S. Ridgway and R. Harrison (London: Academic Press, 1981), 261–72.

131 *eating fish, krill, and crabeater seal pups:* Krause, Goebel, and Kurle, "Leopard Seal Diets," 1–32.

131 *Since 1979, ice habitat in the Western Antarctic Peninsula has de-creased by almost half:* Forcada et al., "Responses of Antarctic Pack-Ice Seals," 40–50.

131 *Shirreff rose sharply between 1998 and 2011:* Krause, Goebel, and Kurle, "Leopard Seal Diets," 1–32.

131 *Since 2010, an average of 70 percent of the pups born have been consumed:* Douglas Krause et al., "The Rapid Population Collapse of a Key Marine Predator in the Northern Antarctic Peninsula Endangers Genetic Diversity and Resilience to Climate Change," *Frontiers in Marine Science* 8 (2022), https://doi.org/10.3389/fmars.2021.796488.

133 *which today accounts for 97 percent of all antarctic fur seals:* A. Grebieniow et al., "Antarctic Fur Seal (*Arctocephalus gazella*) Annual Migration and Temporal Patterns of On-Shore Occurrence of Leucistic Individuals on King George Island," *Polar Biology* 43 (2020): 929–35.

140 *64 percent had experienced sexual harassment and 20 percent had been sexually assaulted:* K. B. H. Clancy et al., "Survey of Academic Field Experiences (SAFE): Trainees Report Harassment and Assault," *PLoS ONE* 9, no. 7 (2014): e102172.

Chapter II: Late January

147 *to serve as a creation myth for the human presence in Antarctica:* E. Glasberg, *Antarctica as Cultural Critique: The Gendered Politics of Scientific Exploration and Climate Change*, 1st ed., Critical

Studies in Gender, Sexuality, and Culture (New York: Palgrave Macmillan, 2012).

148 *"hurled down the penguins . . . dispossessed of the island":* Maddison, *Class and Colonialism.*

148 *a research project run by the country's Ross Sea Region Research and Monitoring Programme:* V. O. van Uitregt et al., "Māori and Antarctica: Ka Mua, Ka Muri Research Report," https://maori antarctica.org/report/, last updated June 14, 2021.

Chapter 13: Mid-February

174 *"We want to be fed with a large wooden spoon":* Alfred Lansing, *Endurance,* 2nd ed. (New York: Carroll, 1999).

Chapter 14: Late February

186 *while still getting additional meals from their parents:* Jefferson Hinke et al., "Divergent Responses of *Pygoscelis* Penguins Reveal Common Environmental Driver," *Oecologia* 153 (2007): 845–55, https://doi.org/10.1007/s00442-007-0781-4.

186 *Since 1994, gentoo penguins have expanded their breeding range thirty-seven miles southward:* R. Herman et al., "Update on the Global Abundance and Distribution of Breeding Gentoo Penguins (*Pygoscelis papua*)," *Polar Biology* 43 (2020): 1947–56.

187 *disproportionately impacted by an increasingly krill-limited system:* Hinke et al., "Divergent Responses," 845–55.

188 *krill fishery is shifting toward being more active in the autumn and winter months:* Nicol, *Curious Life of Krill.*

188 *Overall catch limits are less than 1 percent of the estimated stock of antarctic krill:* S. Hill et al., "Is Current Management of the Antarctic Krill Fishery in the Atlantic Sector of the Southern Ocean Precautionary?," *CCAMLR Science* 23 (2016): 31–51.

188 *where predators, such as penguins and whales, forage:* B. G. Weinstein et al., "Identifying Overlap Between Humpback Whale Foraging Grounds and the Antarctic Krill Fishery," *Biological Conservation* 210 (2017): 184–91; W. Z. Trivelpiece et al., "Variability in Krill Biomass Links Harvesting and Climate Warming to Penguin Population Changes in Antarctica," *Proceedings of the National Academy of Sciences of the United States of America* 108, no. 18 (2011): 7625–28.

188 *at a scale that is much bigger and coarser:* Watters, Hinke, and Reiss, "Long-Term Observations," 2314.

188 *Adélie and chinstrap penguins, both krill dependent, have declined by more than 50 percent:* Trivelpiece et al., "Variability in Krill Biomass," 7625–28.

189 *used in China as a high-value additive in aquaculture feed:* David A. Kroodsma et al., "Tracking the Global Footprint of Fisheries," *Science* 359, no. 6378 (2018): 904–8.

189 *Ten percent of harvested krill are processed into krill oil:* Nicol, *Curious Life of Krill.*

189 *been suggested to have health benefits:* Jian-Ping Yuan et al., "Potential Health-Promoting Effects of Astaxanthin: A High-Value Carotenoid Mostly from Microalgae," *Molecular Nutrition & Food Research* 55, no. 1 (2011): 150–65.

194 *1.55 million square miles:* John Kerry, "On the New Marine Protected Area in Antarctica's Ross Sea," press statement, October 27, 2016, https://2009-2017.state.gov/secretary/remarks/2016/10/263763.htm.

194 *would bring this figure to 39 percent:* M. A. Hindell et al., "Tracking of Marine Predators to Protect Southern Ocean Ecosystems," *Nature* 580 (2020): 87–92.

Chapter 15: Early March

198 *the Zed Islands . . . host a chinstrap colony of twenty thousand pairs:* N. Strycker et al., "A Global Population Assessment of the Chinstrap Penguin (*Pygoscelis antarctica*)," *Scientific Reports* 10 (2020), https://doi.org/10.1038/s41598-020-76479-3.